Fusion
The Energy of the Universe

WHAT IS THE COMPLEMENTARY SCIENCE SERIES?

We hope you enjoy this book. If you would like to read other quality science books with a similar orientation see the order form and reproductions of the front and back covers of other books in the series at the end of this book.

The **Complementary Science Series** is an introductory, interdisciplinary, and relatively inexpensive series of paperbacks for science enthusiasts. The series covers core subjects in chemistry, physics, and biological sciences but often from an interdisciplinary perspective. They are deliberately unburdened by excessive pedagogy, which is distracting to many readers, and avoid the often plodding treatment in many textbooks.

These titles cover topics that are particularly appropriate for self-study although they are often used as complementary texts to supplement standard discussion in textbooks. Many are available as examination copies to professors teaching appropriate courses.

The series was conceived to fill the gaps in the literature between conventional textbooks and monographs by providing real science at an accessible level, with minimal prerequisites so that students at all stages can have expert insight into important and foundational aspects of current scientific thinking.

Many of these titles have strong interdisciplinary appeal and all have a place on the bookshelves of literate laypersons.

Potential authors are invited to contact our editorial office:
j.hayhurst@elsevier.com.
Feedback on the titles is welcome.

Titles in the *Complementary Science Series* are detailed at the end of these pages. A 15% discount is available (to owners of this edition) on other books in this series—see order form at the back of this book.

ELSEVIER
ACADEMIC
PRESS

Physics
Physics in Biology and Medicine, 2nd Edition; Paul Davidovits, 0122048407

Introduction to Relativity; John B. Kogut, 0124175619

Fusion: The Energy of the Universe; Garry McCracken and Peter Stott, 012481851X

Chemistry
The Physical Basis of Chemistry, 2nd Edition; Warren S. Warren, 0127358552

Chemistry Connections: The Chemical Basis of Everyday Phenomena, 2nd Edition; Kerry K. Karukstis and Gerald R. Van Hecke, 0124001513

Fundamentals of Quantum Chemistry; James E. House, 0123567718

Introduction to Quantum Mechanics; S. M. Blinder, 0121060519

Geology
Earth Magnetism; Wallace Hall Campbell, 0121581640

www.books.elsevier.com

Fusion
The Energy of the Universe

Garry McCracken
and
Peter Stott

ELSEVIER
ACADEMIC
PRESS

Amsterdam   Boston   Heidelberg   London   New York   Oxford
Paris   San Diego   San Francisco   Singapore   Sydney   Tokyo

Acquisition Editor: Jeremy Hayhurst
Project Manager: Troy Lilly
Editorial Assistant: Desireé Marr
Marketing Manager: Linda Beattie
Cover Design: Eric DeCicco
Composition: Cepha Imaging Private Limited
Interior Printer: Hing Yip Printing Company

Elsevier Academic Press
200 Wheeler Road, Burlington, MA 01803, USA
525 B Street, Suite 1900, San Diego, California 92101-4495, USA
84 Theobald's Road, London WC1X 8RR, UK

This book is printed on acid-free paper.

Copyright © 2005, Elsevier Inc. All rights reserved.

No part of this publication may be reproduced or transmitted in any form or by any means, electronic or mechanical, including photocopy, recording, or any information storage and retrieval system, without permission in writing from the publisher.

Permissions may be sought directly from Elsevier's Science & Technology Rights Department in Oxford, UK: phone: (+44) 1865 843830, fax: (+44) 1865 853333, e-mail: permissions@elsevier.com.uk. You may also complete your request on-line via the Elsevier homepage (http://elsevier.com), by selecting "Customer Support" and then "Obtaining Permissions."

Library of Congress Cataloging-in-Publication Data

McCracken, G. M.
    Fusion : the energy of the universe / Garry M. McCracken and Peter Stott.
       p. cm. – (Complementary science series)
    Includes bibliographical references and index.
    ISBN 0-12-481851-X (pbk. : acid-free paper)
   1. Nuclear fusion. I. Stott, P. E. (Peter E.) II. Title.
III. Series.
    QC791.M33 2004
    539.7'64–dc22

2004011926

British Library Cataloguing in Publication Data
A catalogue record for this book is available from the British Library

ISBN: 0-12-481851-X

For all information on Elsevier Academic Press Publications visit our Web site at www.books.elsevier.com

Printed in China
04 05 06 07 08 09 9 8 7 6 5 4 3 2 1

To Pamela and Olga,
   Thank you for your encouragement and your patience.

Contents

Foreword   xv

Preface   xvii

Acknowledgments   xix

1 What Is Nuclear Fusion?   1
   1.1 The Alchemists' Dream 1
   1.2 The Sun's Energy 2
   1.3 Can We Use Fusion Energy? 3
   1.4 Man-Made Suns 3
   1.5 The Rest of the Story 4

2 Energy from Mass   7
   2.1 Einstein's Theory 7
   2.2 Building Blocks 8
   2.3 Something Missing 12

3 Fusion in the Sun and Stars   17
   3.1 The Source of the Sun's Energy 17
   3.2 The Solar Furnace 19
   3.3 Gravitational Confinement 20
   3.4 The Formation of Heavier Atoms 24
   3.5 Stars and Supernovae 26

4 Man-Made Fusion   33
   4.1 Down to Earth 33
   4.2 Getting It Together 36
   4.3 Breaking Even 41

5 Magnetic Confinement — 47
5.1 The First Experiments — 47
5.2 Behind Closed Doors — 52
5.3 Opening the Doors — 55
5.4 ZETA — 58
5.5 From Geneva to Novosibirsk — 59

6 The Hydrogen Bomb — 61
6.1 The Background — 61
6.2 The Problems — 63
6.3 Beyond the "Sloyka" — 66

7 Inertial-Confinement Fusion — 69
7.1 Mini-Explosions — 69
7.2 Using Lasers — 72
7.3 Alternative Drivers — 81
7.4 The Future Program — 86

8 False Trails — 87
8.1 Fusion in a Test Tube? — 87
8.2 Bubble Fusion — 91
8.3 Fusion with Mesons — 92

9 Tokamaks — 95
9.1 The Basics — 95
9.2 Instabilities — 96
9.3 Diagnosing the Plasma — 100
9.4 Impurities — 102
9.5 Heating the Plasma — 106

10 From T3 to ITER — 111
10.1 The Big Tokamaks — 111
10.2 Pushing to Peak Performance — 115
10.3 Tritium Operation — 118
10.4 Scaling to a Power Plant — 119
10.5 The Next Step — 123
10.6 ITER — 125

11 Fusion Power Plants — 129
11.1 Early Plans — 129
11.2 Fusion Power Plant Geometry — 129
11.3 Magnetic-Confinement Fusion — 131
11.4 Inertial-Confinement Fusion — 133
11.5 Tritium Breeding — 137

Contents

 11.6 Radiation Damage and Shielding 138
 11.7 Low-Activation Materials 141

12 Why We Need Fusion Energy   **145**
 12.1 World Energy Needs 145
 12.2 The Choice of Fuels 146
 12.3 The Environmental Impact of Fusion Energy 150
 12.4 The Cost of Fusion Energy 152

Epilogue   155

Units   157

Glossary   161

Further Reading   177

Index   181

Technical Summaries

These technical summaries, contained in shaded boxes, are supplements to the main text. They are intended for the more technically minded reader and may be bypassed by the general reader without loss of continuity.

Chapter 2
    2.1   The Mass Spectrograph 11

Chapter 3
    3.1   The Neutrino Problem 21
    3.2   The Carbon Cycle 22
    3.3   Cosmic Microwave Background Radiation 25
    3.4   The Triple Alpha Process 27
    3.5   Heavier Nuclei 28

Chapter 4
    4.1   Source of Deuterium 34
    4.2   Tritium Breeding Reactions 36
    4.3   Conditions for Confinement 44

Chapter 5
    5.1   Magnetic Confinement 49
    5.2   Toroidal Confinement 54
    5.3   Linear Confinement 57

Chapter 7
    7.1   Conditions for Inertial Confinement 71
    7.2   Capsule Compression 72
    7.3   The Laser Principle 74

7.4	Capsule Design	82
7.5	Fast Ignition	83

Chapter 8

8.1	Electrolysis	89

Chapter 9

9.1	Disruptions	98
9.2	Sawteeth	99
9.3	Operating a Tokamak	100
9.4	Temperature Measurement	101
9.5	Sources of Impurities	103
9.6	Impurity Radiation	104
9.7	Production of Neutral Heating Beams	107
9.8	Radio Frequency Heating	108
9.9	L and H Modes	109

Chapter 10

10.1	Operating Limits	116
10.2	Pulse Length and Confinement Time	117
10.3	Understanding Confinement	121
10.4	Empirical Scaling	122

Chapter 11

11.1	Shielding the Superconducting Coils	132
11.2	Power Handling in the Divertor	134
11.3	Driver Efficiency and Target Gain	136
11.4	Tritium Breeding	137
11.5	Choice of the Chemical Form of Lithium	138
11.6	Alternative Fuels	139
11.7	Radiation Damage	140
11.8	Low Activation Materials	142

Foreword

Fusion powers the stars and could in principle provide almost unlimited, environmentally benign, power on Earth. Harnessing fusion has proved to be a much greater scientific and technical challenge than originally hoped. In the early 1970s the great Russian physicist Lev Andreevich Artsimovich wrote that, nevertheless, "thermonuclear [fusion] energy will be ready when mankind needs it." It looks as if he was right and that that time is approaching. This excellent book is therefore very timely.

The theoretical attractions of fusion energy are clear. The raw fuels of a fusion power plant would be water and lithium. The lithium in one laptop computer battery, together with half a bath of water, would generate 200,000 kWh of electricity — as much as 40 tons of coal. Furthermore, a fusion power plant would not produce any atmospheric pollution (greenhouse gases, sulphur dioxide, etc.), thus meeting a requirement that is increasingly demanded by society.

The Joint European Torus (JET), at Culham in the United Kingdom, and the Tokamak Fusion Test Reactor (TFTR), at Princeton in the United States, have produced more than 10 MW (albeit for only a few seconds), showing that fusion can work in practice. The next step will be to construct a power-plant-size device called the International Thermonuclear Experimental Reactor (ITER), which will produce 500 MW for up to 10 minutes, thereby confirming that it is possible to build a full-size fusion power plant. The development of fusion energy is a response to a global need, and it is expected that ITER will be built by a global collaboration.

A major effort is needed to test the materials that will be needed to build fusion plants that are reliable and, hence, economic. If this work is done in parallel with ITER, a prototype fusion power plant could be putting electricity into the grid within 30 years. This is the exciting prospect with which this book concludes.

As early as 1920 it was suggested that fusion could be the source of energy in the stars, and the detailed mechanism was identified in 1938. It was clear by the 1940s that fusion energy could in principle be harnessed on Earth, but early optimism was soon recognized as being (in Artsimovich's words of 1962) "as unfounded as the sinner's hope of entering paradise without passing through purgatory." That purgatory involved identifying the right configuration of magnetic fields to hold a gas at over 100 million degrees Celsius (10 times hotter than the center of the Sun) away from the walls of its container. The solution of this challenging problem — which has been likened to holding a jelly with elastic bands — took a long time, but it has now been found.

Garry McCracken and Peter Stott have had distinguished careers in fusion research. Their book appears at a time when fusion's role as a potential ace of trumps in the energy pack is becoming increasingly recognized. I personally cannot imagine that sometime in the future, fusion energy will not be widely harnessed to the benefit of mankind. The question is when. This important book describes the exciting science of, the fascinating history of, and what is at stake in mankind's quest to harness the energy of the stars.

Chris Llewellyn Smith

(Professor Sir Chris Llewellyn Smith FRS is Director UKAEA Culham Division, Head of the Euratom/UKAEA Fusion Association, and Chairman of the Consultative Committee for Euratom on Fusion. He was Director General of CERN [1994–98]).

Preface

Our aim in writing this book is to answer the frequently asked question "What is nuclear fusion?" In simple terms, *nuclear fusion* is the process in which two light atoms combine to form a heavier atom, in contrast to *nuclear fission* — in which a very heavy atom splits into two or more fragments. Both fusion and fission release energy. Perhaps because of the similarity of the terms, fission and fusion are sometimes confused. Nuclear fission is well known, but in fact nuclear fusion is much more widespread — fusion occurs continuously throughout the universe, and it is the process by which the Sun and the stars release energy and produce new elements from primordial hydrogen. It is a remarkable story.

There has been considerable research effort to use fusion to produce energy on Earth. Fusion would provide an environmentally clean and limitless source of energy. However, to release fusion energy, the fuel has to be heated to unbelievably high temperatures in the region of hundreds of millions of degrees Celsius — hotter in fact than the Sun. The obvious problem is how to contain such very hot fuel — clearly there are no material containers that will withstand such temperatures. There are two alternative ways to solve this problem. The first approach uses magnetic fields to form an insulating layer around the hot fuel. This approach, known as *magnetic confinement*, is now, after 50 years of difficult research, at the stage where a prototype power plant could be built. The second approach is to compress and heat the fuel very quickly so that it burns and the fusion energy is released before the fuel has time to expand. This approach, known as *inertial confinement*, is still at the stage where the scientific feasibility remains to be demonstrated.

In this book we present the complete story of fusion, starting with the development of the basic scientific ideas that led to the understanding of the role of fusion in the Sun and stars. We explain the processes of hydrogen burning in the Sun and the production of heavier elements in stars and supernovae. The development of fusion as a source of energy on Earth by both the magnetic and inertial confinement approaches is discussed in detail from the scientific beginnings to the construction of a fusion power plant. We briefly explain the principles of the hydrogen bomb and also review various false trails to fusion energy. The final chapter looks at fusion in the context of world energy needs.

The book has been structured to appeal to a wide readership. In particular we hope it will appeal to readers with a general interest in science but little scientific background as well as to students who may find it useful as a supplement to more formal textbooks. The main text has been written with the minimum of scientific jargon and equations and emphasizes a simple and intuitive explanation of the scientific ideas. Additional material and more technical detail is included in the form of shaded "boxes" that will help the more serious student to understand some of the underlying physics and to progress to more advanced literature. However, these boxes are not essential reading, and we encourage the nonscientist to bypass them — the main text contains all that is needed to understand the story of fusion. We have tried to present the excitement of the scientific discoveries and to include brief portraits of some of the famous scientists who have been involved.

November 2004

Acknowledgments

In the course of writing this book we have drawn on the vast volume of published material relating to fusion in scientific journals and elsewhere as well as on unpublished material and discussions with our colleagues. We have tried to give an accurate and balanced account of the development of fusion research that reflects the relative importance of the various lines that have been pursued and gives credit to the contributions from the many institutions in the countries that have engaged themselves in fusion research. However, inevitably there will be issues, topics, and contributions that some readers might feel deserved more detailed treatment.

We would like to thank all of our colleagues who have helped and advised us in many ways. In particular we are indebted to John Wesson, Jim Hastie, Bruce Lipschitz, Peter Stangeby, Spencer Pitcher, and Stephen Pitcher, who took a great deal of time and trouble to read early drafts of the book and who gave constructive criticism and valuable suggestions for its improvement. We are grateful also to Chris Carpenter, Jes Christiansen, Geoff Cordey, Richard Dendy, John Lawson, Ramon Leeper, Kanetada Nagamine, Peter Norreys, Neil Taylor, Fritz Wagner, David Ward, Alan Wootton, and many others who have helped us to check specific points of detail, who generously provided figures, and who assisted in other ways. A special thanks is due to Jeremy Hayhurst and Troy Lilly and their colleagues at Elsevier Academic Press Publishing for their patience and encouragement.

The contents of this book and the views expressed herein are the sole responsibility of the authors and do not necessarily represent the views of the European Commission or the European Fusion Program.

The authors thank the following organizations and individuals for granting permission to use the following figures: EFDA-JET (for Figs. 1.2, 2.5, 3.8, 4.4, 4.5, 4.6, 4.8, 5.2, 5.4, 9.2, 9.3, 9.8, 10.1, 10.3, 10.5, 10.6, 10.7, 10.8, and 11.1); UKAEA Culham Laboratory (for Figs. 5.3, 5.6, 5.7, 5.8, 11.2, 11.5, and 12.3, and also for permission to use the archive photographs in Figs. 5.1, 5.9, 9.1, and 10.2); The University of California/Lawrence Livermore National Laboratory (for Figs. 7.6, 7.7, 7.8, 7.9, 7.10, and 7.14); Sandia National Laboratories (for Figs. 7.12 and 7.13); Max-Planck-Institut für Plasmaphysik Garching (for Fig. 9.6); Princeton Plasma Physics Laboratory (for Fig. 10.4); ITER (for Fig. 10.10); John Lawson (for Fig. 4.7); National Library of Medicine (for Fig. 1.1); Evelyn Einstein (for Fig. 2.1—used by permission); The Royal Society (for Fig. 2.3); Master and Fellows of Trinity College, Cambridge (for Fig. 2.4—used by permission); NASA image from Solarviews.com (for Fig. 3.3); P Emilio Segrè Visual Archives (for Figs. 3.4 and 5.5); Anglo-Australian Observatory/David Malin Images (for Figs. 3.6 and 3.7); Teller Family Archive as appeared in *Edward Teller* by Stanley A. Blumberg and Louis G. Panos, Scribner's Sons, 1990 (for Fig. 6.1); VNIIEF Museum and Archive, Courtesy AIP Emilion Segrè Visual Archives, Physics Today Collection (for Fig. 6.4); The Nobel Foundation (for Fig. 7.5—used by permission); Fig. 3.1 originally appeared in *The Life of William Thomson*, Macmillan, 1910. Copyright for these materials remains with the original copyright holders.

Every effort has been made to contact all copyright holders. The authors would be pleased to make good in any future editions any errors or omissions that are brought to their attention.

Some figures have been redrawn or modified by the authors from the originals. Figure 3.5 is adapted from *Nucleosynthesis and Chemical Evolution of Galaxies* by B.E.J. Pagel (Cambridge University Press, Cambridge, UK, 1997); Fig. 6.2 is adapted from *Dark Sun: The making of the hydrogen bomb*, by R. Rhodes (Simon and Schuster, New York, NY, 1996); Figs. 7.1, 7.2, and 7.11 are adapted from J. Lindl, *Physics of Plasmas*, **2** (1995) 3939; Fig. 8.1 is adapted from *Too Hot to Handle: The Story of the Race for Cold Fusion* by F. Close (W. H. Allen Publishing, London, 1990; Fig. 8.2 is adapted from K. Ishida *et al*, *J. Phys.* **G 29** (2003) 2043; Fig. 9.4 is adapted from an original drawing by General Atomics; Fig. 9.5 is adapted from *L'Energie des Etoiles, La Fusion Nucleaire Controlée* by P-H Rebut (Editions Odile Jacob Paris, 1999); Fig. 12.1 is adapted from the World Energy Council Report, *Energy for Tomorrow's World: The realities, the real options and the agenda for achievement* (St. Martins Press, New York, NY 1993); Fig. 12.2 is adapted from *Key World Energy Statistics from the IEA: 2003 Edition* (IEA: Paris, 2004). We are particularly grateful to Stuart Morris and his staff, who drew or adapted many of the figures specifically for this book.

Chapter 1

What Is Nuclear Fusion?

1.1 The Alchemists' Dream

In the Middle Ages, the alchemists' dream was to turn lead into gold. The only means of tackling this problem were essentially chemical ones, and these were doomed to failure. During the 19th century the science of chemistry made enormous advances, and it became clear that lead and gold are different elements that cannot be changed into each other by chemical processes. However, the discovery of radioactivity at the very end of the 19th century led to the realization that sometimes elements do change spontaneously (or *transmute*) into other elements. Later, scientists discovered how to use high-energy particles, either from radioactive sources or accelerated in the powerful new tools of physics that were developed in the 20th century, to induce artificial transmutations in a wide range of elements. In particular, it became possible to split atoms (the process known as *nuclear fission*) or to combine them together (the process known as *nuclear fusion*). The alchemists (Fig. 1.1) did not understand that their quest was impossible with the tools they had at their disposal, but in one sense it could be said that they were the first people to search for nuclear transmutation.

What the alchemists did not realize was that nuclear transmutation was occurring before their very eyes, in the Sun and in all the stars of their night sky. The processes in the Sun and stars, especially the energy source that had sustained their enormous output for eons, had long baffled scientists. Only in the early 20th century was it realized that nuclear fusion is the energy source that runs the universe and that simultaneously it is the mechanism responsible for creating all the different chemical elements around us.

Figure 1.1 ▸ An alchemist in search of the secret that would change lead into gold. Because alchemists had only chemical processes available, they had no hope of making the nuclear transformation required. (An engraving from a painting by David Teniers the younger, 1610–1690.)

1.2 The Sun's Energy

The realization that the energy radiated by the Sun and stars is due to nuclear fusion followed three main steps in the development of science. The first was Albert Einstein's famous deduction in 1905 that mass can be converted into energy. The second step came a little over 10 years later with Francis Aston's precision measurements of atomic masses, which showed that the total mass of four hydrogen atoms is slightly larger than the mass of one helium atom. These two key results led Arthur Eddington and others, around 1920, to propose that mass could be turned into energy in the Sun and the stars if four hydrogen atoms combine to form a single helium atom. The only serious problem with this model was that, according to classical physics, the Sun was not hot enough for nuclear fusion to take place. It was only after *quantum mechanics* was developed in the late 1920s that a complete understanding of the physics of nuclear fusion became possible.

Having answered the question as to where the energy of the universe comes from, physicists started to ask how the different atoms arose. Again fusion was the answer. The fusion of hydrogen to form helium is just the start of a long and complex chain. It was later shown that three helium atoms can combine to form a carbon atom and that all the heavier elements are formed in a series of more and more complicated reactions. Nuclear physicists played a key role in reaching these conclusions. By studying the different nuclear reactions in laboratory accelerators, they were able to deduce the most probable reactions under different conditions. By relating these data to the astrophysicists' models of the stars, a consistent picture of the life cycles of the stars was built up and the processes that give rise to all the different atoms in the universe were discovered.

1.3 Can We Use Fusion Energy?

When fusion was identified as the energy source of the Sun and the stars, it was natural to ask whether the process of turning mass into energy could be demonstrated on Earth and, if so, whether it could be put to use for man's benefit. Ernest Rutherford, the famous physicist and discoverer of the structure of the atom, made this infamous statement to the British Association for the Advancement of Science in 1933: "We cannot control atomic energy to an extent that would be of any use commercially, and I believe we are not ever likely to do so." It was one of the few times when his judgment proved wanting. Not everybody shared Rutherford's view; H. G. Wells had predicted the use of nuclear energy in a novel published in 1914.[1]

The possibility of turning nuclear mass into energy became very much more real in 1939 when Otto Hahn and Fritz Strassman demonstrated that the uranium atom could be split by bombarding uranium with neutrons, with the release of a large amount of energy. This was fission. The story of the development of the fission chain reaction, fission reactors, and the atom bomb has been recounted many times. The development of the hydrogen bomb and the quest for fusion energy proved to be more difficult. There is a good reason for this. The uranium atom splits when bombarded with neutrons. Neutrons, so called because they have no electric charge, can easily penetrate the core of a uranium atom, causing it to become unstable and to split. For fusion to occur, two hydrogen atoms have to get so close to each other that their cores can merge; but these cores carry strong electric charges that hold them apart. The atoms have to be hurled together with sufficiently high energy to make them fuse.

1.4 Man-Made Suns

The fusion reaction was well understood by scientists making the first atomic (fission) bomb in the Manhattan Project. However, although the possibility that

---

[1] *Atomic energy* and *nuclear energy* are the same thing.

fusion could be developed as a source of energy was undoubtedly discussed, no practical plans were put forward. Despite the obvious technical difficulties, the idea of exploiting fusion energy in a controlled manner was seriously considered shortly after World War II, and research was started in the UK at Liverpool, Oxford, and London universities. One of the principal proponents was George Thomson, the Nobel Prize–winning physicist and son of J. J. Thomson, the discoverer of the electron. The general approach was to try to heat hydrogen gas to a high temperature so that the colliding atoms have sufficient energy to fuse together. By using a magnetic field to confine the hot fuel, it was thought that it should be possible to allow adequate time for the fusion reactions to occur. Fusion research was taken up in the UK, the US, and the Soviet Union under secret programs in the 1950s and subsequently, after being declassified in 1958, in many of the technically advanced countries of the world. The most promising reaction is that between the two rare forms of hydrogen, called *deuterium* and *tritium*. Deuterium is present naturally in water and is therefore readily available. Tritium is not available naturally and has to be produced *in situ* in the power plant. This can be done by using the products of the fusion reaction to interact with the light metal lithium in a layer surrounding the reaction chamber in a *breeding cycle*. Thus the basic fuels for nuclear fusion are lithium and water, both readily and widely available. Most of the energy is released as heat that can be extracted and used to make steam and drive turbines, as in any conventional power plant. A schematic diagram of the proposed arrangement is shown in Fig. 1.2. The problem of heating and containing the hot fuel with magnetic fields turned out to be much more difficult than at first envisaged.

However, research on the peaceful use of fusion energy was overtaken in a dramatic way with the explosion of the hydrogen bomb in 1952. This stimulated a second approach to controlled fusion, based on the concept of heating the fuel to a sufficiently high temperature very quickly before it has time to escape. The invention of the laser in 1960 provided a possible way to do this; lasers can focus intense bursts of energy onto small targets. The idea is to rapidly heat and compress small fuel pellets or capsules in a series of mini-explosions. This is called *inertial confinement* because the fusion fuel is confined only by its own inertia. Initially the expertise was limited to those countries that already had nuclear weapons, and some details still remain a close secret, although other countries have now taken it up for purely peaceful purposes. Apart from the heating and confinement of the fuel, the method of converting fusion energy into electricity will be very similar to that envisaged for magnetic confinement.

1.5 The Rest of the Story

The considerable scientific and technical difficulties encountered by the magnetic- and inertial-confinement approaches have caused these programs to stretch over many years. The quest for fusion has proved to be one of the most difficult

Section 1.5 The Rest of the Story

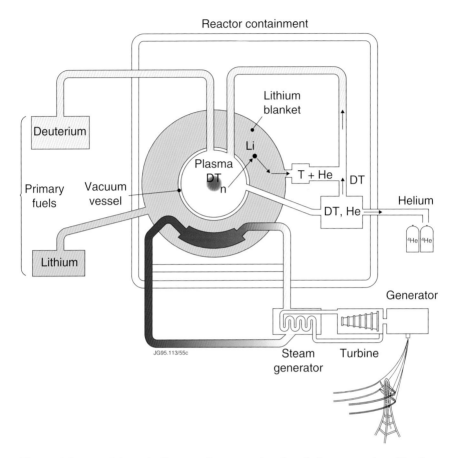

Figure 1.2 ▶ Schematic diagram of a proposed nuclear fusion power plant. The deuterium and tritium fuel burns at a very high temperature in the central reaction chamber. The energy is released as charged particles, neutrons, and radiation and it is absorbed in a lithium blanket surrounding the reaction chamber. The neutrons convert the lithium into tritium fuel. A conventional steam-generating plant is used to convert the nuclear energy to electricity. The waste product from the nuclear reaction is helium.

challenges faced by scientists. After many years, the scientific feasibility of thermonuclear fusion via the magnetic-confinement route has been demonstrated, and the next generation of inertial-confinement experiments is expected to reach a similar position. Developing the technology and translating these scientific achievements into power plants that are economically viable will be a major step that will require much additional time and effort. Some have hoped that they could find easy ways to the rewards offered by fusion energy. This line of thinking has led to many blind alleys and even to several false claims of success, the most widely publicized being the so-called "cold fusion" discoveries that are described in Chapter 8.

▶ Chapter 2

Energy from Mass

2.1 Einstein's Theory

Energy is something with which everyone is familiar. It appears in many different forms, including electricity, light, heat, chemical energy, and motional (or *kinetic*) energy. An important scientific discovery in the 19th century was that *energy is conserved*. This means that energy can be converted from one form to another but that the total amount of energy must stay the same. *Mass* is also very familiar, though sometimes it is referred to, rather inaccurately, as weight. On the Earth's surface, mass and weight are often thought of as being the same thing, and they do use the same units — something that weighs 1 kilogram has a mass of 1 kilogram — but strictly speaking *weight* is the force that a mass experiences in the Earth's gravity. An object always has the same mass, even though in outer space it might appear to be weightless. Mass, like energy, is conserved.

The extraordinary idea that mass and energy are equivalent was proposed by Albert Einstein (Fig. 2.1) in a brief three-page paper published in 1905. It was written by a young man who was virtually unknown in the scientific world. His paper on the equivalence of mass and energy followed soon after three seminal papers — on the photoelectric effect, on Brownian motion, and on special relativity — all published in the same year. Henri Becquerel had discovered radioactivity 10 years previously. Using simple equations and the application of the laws of conservation of energy and momentum, Einstein argued that the atom left after a radioactive decay event had emitted energy in the form of radiation must be less massive than the original atom. From this analysis he deduced that "If a body gives off the energy E in the form of radiation, its mass diminishes by E/c^2." He went on to say, "It is not impossible that with bodies whose energy content is variable to a high degree (e.g., radium salts) the theory may be successfully put to the test."

Figure 2.1 ▶ Wedding photograph of Maria Maric and Albert Einstein, January 1903. Einstein had graduated in 1901 and had made a number of applications for academic jobs, without success. He eventually got a job as technical expert, third class, in the Swiss patent office in Berne, which meant that he had to do all his research in his spare time.

Einstein's deduction is more commonly written as $E = mc^2$, probably the most famous equation in physics. It states that mass is another form of energy and that energy equals mass multiplied by the velocity of light squared. Although it took a long time to get experimental proof of this entirely theoretical prediction, we now know that it was one of the most significant advances ever made in science.

2.2 Building Blocks

To see how Einstein's theory led to the concept of fusion energy we need to go back to the middle of the 19th century. As the science of chemistry developed, it became clear that everything is built up from a relatively small number of basic components called *elements*. At that time about 50 elements had been identified, but we now know that there are around 100. As information accumulated about the different elements it became apparent that there were groups of them with similar properties. However, it was not clear how these were related to each other until the *Periodic Table* was proposed by the Russian chemist Dmitri Mendeleev. In 1869 he published a table in which the elements were arranged in rows, with the lightest elements, such as hydrogen, in the top row and the heaviest in the bottom row.

Section 2.2 **Building Blocks**

Elements with similar physical and chemical properties were placed in the same vertical columns. The table was initially imperfect, mainly because of inaccuracies in the data and because some elements had not yet been discovered. In fact, gaps in Mendeleev's table stimulated the search for and the discovery of new elements.

Each element consists of tiny units called *atoms*. Ernest Rutherford deduced in 1911 that atoms have a heavy core called the *nucleus* that has a positive electric charge. A cloud of lighter particles called *electrons* with a negative electric charge surrounds the nucleus. The negative electric charges of the electrons and the positive charge of the nucleus balance each other so that the atom overall has no net electric charge. The number of positive charges and electrons is different for each element, and this determines the element's chemical properties and its position in Mendeleev's table. Hydrogen is the simplest element, with just one electron in each atom; helium is next, with two electrons; lithium has three; and so on down to uranium, which, with 92 electrons, is the heaviest naturally occurring element. Schematic diagrams of the structure of the atoms of hydrogen and helium are shown in Fig. 2.2.

The chemists developed skilled techniques to measure the average mass of the atoms of each element — the *atomic mass* (this is also known as the *atomic weight*). Many elements were found to have atomic masses that were close to being simple multiples of the atomic mass of hydrogen, and this suggested that,

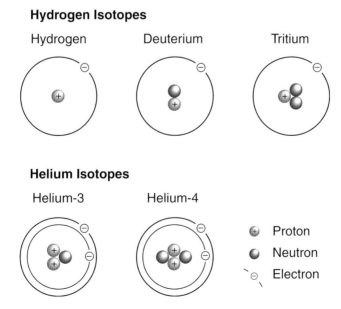

Figure 2.2 ▶ Structure of the different atoms of hydrogen and helium. Atoms with the same number of protons and different numbers of neutrons are known as *isotopes* of the same element.

in some way that was not understood at that time, hydrogen might be a sort of building block for the heavier elements. To take some common examples, the atomic mass of carbon is approximately 12 times that of hydrogen, and the atomic mass of oxygen is 16 times that of hydrogen. There were some puzzling cases, however, that did not fit the general pattern. For example, repeated measurements of the atomic mass of chlorine gave a value of 35.5 times that of hydrogen.

The next significant step in the story was the direct measurement of the masses of individual atoms. During the period 1918–1920 at Cambridge University, UK, Francis Aston (Fig. 2.3) built an instrument (Box 2.1) that could do this. Having studied chemistry at Birmingham, Aston had become interested in passing currents through gases in low-pressure discharge tubes. In 1910 he was invited to the Cavendish laboratory at Cambridge by J. J. Thomson, who was studying positive rays, also by using discharge tubes. Aston helped Thomson to set up an apparatus for measuring the mass-to-charge ratio of the positive species in the discharge.

Figure 2.3 ▶ Francis Aston, 1877–1945, Nobel Laureate in Chemistry 1922. He started his scientific career by setting up a laboratory in a barn at his parents' home while still a schoolboy.

BOX 2.1 The Mass Spectrograph

The Aston mass spectrograph was an important development in the study of atomic masses. Starting by ionizing atoms either in an electric discharge or by an electron beam, a beam of ions is produced that is accelerated in an electric field to a fixed energy, eV, determined by the equation

$$\frac{1}{2}mv^2 = eV$$

where m, v, and e are the mass, velocity, and charge of the ions and V is the voltage through which the ions are accelerated.

The ions then pass into a uniform magnetic field, which exerts a force on them at right angles to the direction of the field and to the direction of the ion. The magnetic field provides the centripetal force on the ions, forcing them to follow a circular path whose radius is given by the equation

$$mv^2/r = Bev$$

Because all the ions have the same energy, the radius r of their circular path depends on their mass-to-charge ratio. The ions are thus dispersed spatially, rather as light is dispersed by a prism. One of the principal advantages of the geometry chosen by Aston is that the ions with the same ratio of mass to charge are spatially focused at the detector, thus optimizing the efficiency with which the ions are collected.

Many variations of the mass spectrograph (using electrical detection it is known as the *mass spectrometer*) have been developed and are widely used for routine analysis of all types of samples. One interesting application is its use for archaeological dating by measuring the ratio of the abundances of two isotopes of an element. If one isotope is radioactive, the age of a sample can be deduced. This analysis is often applied to ^{14}C and to the rubidium isotopes ^{85}Rb and ^{87}Rb, but other elements can be used, depending on the age of the sample being analyzed.

After World War I, Aston returned to Cambridge and started to measure the mass of atoms by a new method, and this was a great improvement on the Thomson apparatus. He subjected the atoms to an electric discharge, which removed one or more of their electrons. This left the nucleus surrounded with a depleted number of electrons and thus with a net positive electric charge — this is known as an *ion*. These ions were accelerated by an electric field to a known energy and then passed through a magnetic field. By measuring the amount by which they were deflected in the magnetic field, Aston was able to determine the mass of the atoms. The instrument was dubbed a *mass spectrograph* because the beams of ions were dispersed into a spectrum in a similar way that a prism disperses light. Aston was a brilliant experimentalist with an obsession for accuracy. Gradually he developed greater and greater precision until he was able to determine the mass of an atom to an accuracy of better than one part in a thousand. These precision measurements yielded a number of entirely unexpected results. It is a good example of pure scientific curiosity leading eventually to valuable practical information.

Aston found that some atoms that are chemically identical could have different masses. This resolved the puzzle about the atomic weight of chlorine. There are two types of chlorine atom; one type is about 35 times heavier than hydrogen, the other about 37 times heavier. The relative abundance of the two types (75% have mass 35 and 25% have mass 37) gives an average of 35.5 — in agreement with the chemically measured atomic mass. Likewise, Aston found that the mass of a small percentage (about 0.016%) of hydrogen atoms is almost double that of the majority. Atoms with different masses but the same chemical properties are called *isotopes*.

The reason for the difference in mass between isotopes of the same element was not understood until 1932, when James Chadwick discovered the neutron. It was then realized that the nucleus contains two types of atomic particle: *protons*, with a single unit of positive electric charge, and *neutrons*, with no electric charge. The number of protons equals the number of electrons, so an atom is overall electrically neutral. All isotopes of the same element have the same number of protons and the same number of electrons, so their chemical properties are identical. The number of neutrons can vary. For example, chlorine always has 17 protons, but one isotope has 18 neutrons and the other has 20. Likewise the nucleus of the most common isotope of hydrogen consists of a single proton; the heavier forms, *deuterium* and *tritium*, have one proton with one and two neutrons, respectively, as shown in Fig. 2.2. Protons and neutrons have very similar masses (the mass of a neutron is 1.00138 times the mass of a proton), but electrons are much lighter (a proton is about 2000 times the mass of an electron). The total number of protons and neutrons therefore determines the overall mass of the atom.

2.3 Something Missing

The most surprising result from Aston's work was that the masses of individual isotopes are not *exactly* multiples of the mass of the most common isotope of hydrogen; they are consistently very slightly lighter than expected. Aston had defined his own scale of atomic mass by assigning a value of precisely 4 to helium. On this scale, the mass of the light isotope of hydrogen is 1.008, so the mass of a helium atom is only 3.97 times rather than exactly 4 times the mass of a hydrogen atom. The difference is small, but Aston's reputation for accuracy was such that the scientific world was quickly convinced by his results.

The significance of this result was quickly recognized by a number of people. One was Arthur Eddington (Fig. 2.4), now considered to be the most distinguished

Section 2.3 **Something Missing**

Figure 2.4 ▶   Arthur Eddington, 1882–1944, from the drawing by Augustus John.

astrophysicist of his generation. He made the following remarkably prescient statement at the British Association for Advancement of Science meeting in Cardiff in 1920, only a few months after Aston had published his results.

> Aston has further shown conclusively that the mass of the helium atom is less than the sum of the masses of the four hydrogen atoms which enter into it and in this at least the chemists agree with him. There is a loss of mass in the synthesis amounting to 1 part in 120, the atomic weight of hydrogen being 1.008 and that of helium just 4.00.... Now mass cannot be annihilated and the deficit can only represent the mass of the electrical energy liberated when helium is made out of hydrogen. If 5% of a star's mass consists initially of hydrogen atoms, which are gradually being combined to form more complex elements, the total heat liberated will more than suffice for our demands, and we need look no further for the source of a star's energy.

> If, indeed, the subatomic energy in the stars is being freely used to maintain their furnaces, it seems to bring a little nearer to fulfillment our dream of controlling this latent power for the well-being of the human race — or for its suicide.

Eddington had realized that there would be a mass loss if four hydrogen atoms combined to form a single helium atom. Einstein's equivalence of mass and energy led directly to the suggestion that this could be the long-sought process that produces the energy in the stars! It was an inspired guess, all the more remarkable because the structure of the nucleus and the mechanisms of these reactions were not fully understood. Moreover, it was thought at that time that there was very little hydrogen in the Sun, which accounts for Eddington's assumption that only 5% of a star's mass might be hydrogen. It was later shown in fact that stars are composed almost entirely of hydrogen.

In fact, according to the classical laws of physics, the processes envisaged by Eddington would require much higher temperatures than exist in the Sun. Fortunately, a new development in physics known as *quantum mechanics* soon provided the answer and showed that fusion can take place at the temperatures estimated to occur in the Sun. The whole sequence of processes that allows stars to emit energy over billions of years was explained in detail by George Gamow, by Robert Atkinson and Fritz Houtermans in 1928, and by Hans Bethe in 1938.

The question as to who first had the idea that fusion of hydrogen into helium was the source of the Sun's energy led to some bitter disputes, particularly between Eddington and James Jeans. Each thought they had priority, and they were on bad terms for many years as a result of the dispute.

As the techniques of mass spectroscopy were refined and made increasingly more accurate, detailed measurements were made on every isotope of every element. It was realized that many isotopes are lighter than would be expected by simply adding up the masses of the component parts of their nuclei — the protons and neutrons. Looked at in a slightly different way, each proton or neutron when combined into a nucleus has slightly less mass than when it exists as a free particle. The difference in mass per nuclear particle is called the *mass defect*, and, when multiplied by the velocity of light squared, it represents the amount of energy associated with the forces that hold the nucleus together.

These data are usually plotted in the form of a graph of the energy equivalent of the mass defect plotted against the total number of protons and neutrons in the nucleus (the atomic mass). A modern version of this plot is shown in Fig. 2.5. While there are some irregularities in the curve at the left-hand side, for the lightest isotopes, most of the curve is remarkably smooth. The most important feature is the minimum around mass number 56. Atoms in this range are the most stable. Atoms to either side have excess mass that can be released in the form of energy by moving toward the middle of the curve, that is, if two lighter atoms join to form a heavier one (this is *fusion*) or a very heavy atom splits to form lighter fragments (this is *fission*).

Section 2.3 ▶ **Something Missing**   **15**

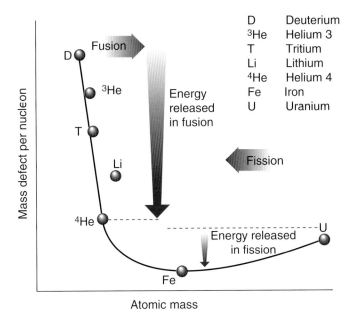

Figure 2.5 ▶   The energy equivalent of the mass defect per nucleon of the elements plotted as a function of their atomic mass. The most stable elements, around iron and nickel, have the lowest energy. The amount of energy released when one atom is transmuted to another, either by fusing light atoms or by splitting heavy ones, is equal to the difference in their masses.

It turns out that splitting the heavy atoms is very much the easier task, but the discovery of how it can be done was quite accidental. After the neutron had been discovered, it occurred to a number of groups to bombard uranium, the heaviest naturally occurring element with an atomic mass of about 238, with neutrons in order to try to make even heavier *transuranic elements*. The amount of any new element was expected to be exceedingly small, and very sensitive detection techniques were required. Some genuine transuranic elements were detected, but there were some reaction products that did not fit the expectations. In 1939 Otto Hahn and Fritz Strassman performed a series of experiments that showed conclusively that these unexplained products were actually isotopes of barium and lanthanum that have mass numbers of 139 and 140, respectively, roughly half the mass of the uranium target nuclei. The only possible explanation was that the neutron bombardment of the uranium had induced fission in the uranium nucleus, causing it to split into two approximately equal parts. Moreover it turned out that additional neutrons were released in the process. It was quickly realized that these neutrons could in principle induce further fission reactions, leading to a chain reaction. This led to the building of the first atomic reactor by Enrico Fermi in Chicago in 1943 and the development of the atomic bomb in Los Alamos.

► Chapter 3

Fusion in the Sun and Stars

3.1 The Source of the Sun's Energy

At the beginning of the 20th century there was no convincing explanation for the enormous amount of energy radiated by the Sun. Although physics had made major advances during the previous century and many people thought that there was little of the physical sciences left to be discovered, they could not explain how the Sun could continue to release energy, apparently indefinitely. The law of energy conservation requires that there be an internal energy source equal to that radiated from the Sun's surface. The only substantial sources of energy known at that time were wood and coal. Knowing the mass of the Sun and the rate at which it radiated energy, it was easy to show that if the Sun had started off as a solid lump of coal it would have burnt out in less than 2000 years. It was clear that this was much too short—the Sun had to be older than the Earth, and the Earth was known to be older than 2000 years—but just how old was the Earth?

Early in the 19th century most geologists had believed that the Earth might be indefinitely old. This idea was disputed by the distinguished physicist William Thomson, who later became Lord Kelvin (Fig. 3.1). His interest in this topic began in 1844 while he was still a Cambridge undergraduate. It was a topic to which he returned repeatedly and that drew him into conflict with other scientists, such as John Tyndall, Thomas Huxley, and Charles Darwin. To evaluate the age of the Earth, Kelvin tried to calculate how long it had taken the planet to cool from an initial molten state to its current temperature. In 1862 he estimated the Earth to be 100 million years old. To the chagrin of the biologists, Kelvin's calculations for the age of the Earth did not allow enough time for evolution to occur. Over the next four decades, geologists, paleontologists, evolutionary biologists, and physicists joined in a protracted debate about the age of the Earth. During this time Kelvin revised his figure down to between

Figure 3.1 ▶ William Thomson, later Lord Kelvin (1824–1907). Kelvin was one of the pioneers of modern physics, developing thermodynamics. He had a great interest in practical matters and helped to lay the first transatlantic telegraph cable.

20 million and 40 million years. The geologists tried to make quantitative estimates based on the time required for the deposition of rock formations or the time required to erode them, and they concluded that the Earth must be much older than Kelvin's values. However, too many unknown factors were required for such calculations, and they were generally considered unreliable. In the first edition of his book, *The Origin of Species*, Charles Darwin calculated the age of the Earth to be 300 million years, based on the time estimated to erode the Weald, a valley between the North and South Downs in southern England. This was subjected to so much criticism that Darwin withdrew this argument from subsequent editions.

The discrepancy between the estimates was not resolved until the beginning of the 20th century, when Ernest Rutherford realized that radioactivity (discovered by Henri Becquerel in 1896, well after Kelvin had made his calculations) provides the Earth with an internal source of heat that slows down the cooling. This process makes the Earth older than was originally envisaged; current estimates suggest that our planet is at least 4.6 billion years old. Radioactivity, as well as providing the additional source of heat, provides an accurate way of measuring the age of the Earth by comparing the amounts of radioactive minerals in the rocks. The age of the Earth put a lower limit on the age of the Sun and renewed the debate about the source of the Sun's energy — what was the mechanism that could sustain the Sun's output for such a long period of time. It was not until the 1920s, when Eddington made his deduction that fusion of hydrogen was the most likely energy source, and later, when quantum theory was developed, that a consistent explanation became possible.

3.2 The Solar Furnace

Hydrogen and helium are by far the most common elements in the universe and together account for about 98% of all known matter. There is no significant amount of hydrogen or helium in the gaseous state on Earth (or on Mars or Venus) because the gravity of small planets is too weak to keep these light atoms attached; they simply escape into outer space. All of the Earth's hydrogen is combined with oxygen as water, with carbon as hydrocarbons, or with other elements in the rocks. However, the Sun, whose gravity is much stronger, consists almost entirely of hydrogen. The presence of hydrogen in the Sun and the stars can be measured directly from spectroscopic observations, since every atom emits light with characteristic *wavelengths* (or colors) that uniquely identify it.

Although there is plenty of hydrogen in the Sun for nuclear fusion, how can we know that conditions are right for fusion to occur? The temperature and density of the Sun can be determined by a combination of experimental observations using spectroscopy and by theoretical calculations. The most likely fusion reactions can be deduced from studies of nuclear reactions in the laboratory, using particle accelerators. The energy release in the Sun involves the conversion of four protons into a helium nucleus. However, this does not happen in a single step. First, two protons combine to form a nucleus of the heavy isotope of hydrogen known as *deuterium*. The deuterium nucleus then combines with another proton to form the light helium isotope known as *helium-3*. Finally two helium-3 nuclei combine to form *helium-4*, releasing two protons in the process. Overall, four protons are converted into one helium nucleus. Energy is released because the helium nucleus has slightly less mass than the original four protons from which it was formed, as discussed in Chapter 2. The structure of the different nuclei was illustrated in Fig. 2.2. The reactions are shown schematically in Fig. 3.2, with the original nuclei on the left-hand side and the products on the right. Energy is released in each stage of the reactions. The total amount of energy released for each conversion of four hydrogen nuclei into a helium nucleus is about 10 million times more than is produced by the chemical reaction when hydrogen combines with oxygen and burns to form water. This enormous difference between the energy released by nuclear reactions compared to chemical reactions explains why fusion can sustain the Sun for billions of years. It is about 10 million times longer in fact than the estimate of a few thousand years that was obtained when the Sun was considered to be a lump of coal.

The energy has to be transported from the Sun's interior core to the surface. This is quite a slow process, and it takes about a million years for the energy to get out. The Sun's surface is cooler than the core and the energy is radiated into space as the heat and light that we observe directly. Under standard conditions, the solar power falling on the Earth is about 1.4 kilowatts per square meter (kW m^{-2}).

The first stage of the reactions just described (see also Box 3.1) is known to nuclear physicists as a *weak interaction*. The process is very slow, and this sets the pace for the conversion to helium. It takes many hundreds of millions of years for two protons to fuse together. This turns out to be rather fortunate. If the fusion

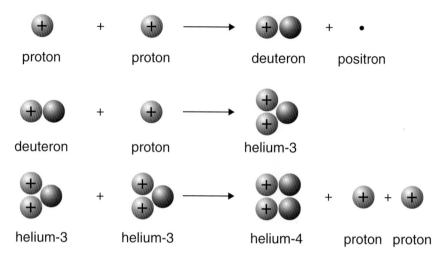

Figure 3.2 ▶ The three reactions that convert hydrogen to helium in the Sun. The overall charge balance is conserved through emission of a *positron*, a particle with the same mass and properties as an electron but with positive electric charge.

reaction took place too quickly, then the Sun would have burned out long before life on Earth had a chance to evolve. From our knowledge of the nuclear reaction rates and of the amount of initial hydrogen, it is estimated that the time to use up all the hydrogen is about 10 billion years. From radioactive dating of meteorites it is estimated that the age of the solar system is 4.6 billion years. Assuming the Sun is the same age as the meteorites, then it is approximately halfway through its life cycle. For comparison, the most recent estimate of the age of the universe is about 13.7 billion years.

The proton–proton reaction tends to dominate in stars that are the size of our Sun or smaller. However, in larger stars there is another reaction cycle, involving reactions with a carbon nucleus (see Box 3.2), by which protons can be converted into helium nuclei.

3.3 Gravitational Confinement

The hydrogen in the Sun's core is compressed to very high density, roughly 10 times denser than lead. But the Sun's core is not solid — it is kept in an ionized, or *plasma*, state by the high temperature. This combination of high density and high temperature exerts an enormous outward pressure that is about 400 billion (4×10^{11}) times larger than the atmospheric pressure at the Earth's surface.

An inward force must balance this enormous outward pressure in order to prevent the Sun from expanding. Gravity provides this force in the Sun and stars, and it compresses the Sun into the most compact shape possible, a sphere. At each layer

BOX 3.1 The Neutrino Problem

The first stage of the chain of reactions in the Sun—that between two protons—also releases a particle called a *neutrino* (ν):

$$p + p \rightarrow D + {}^+e + \nu$$

Neutrinos were first predicted by Enrico Fermi, to explain a discrepancy in the energy measured in the process known as *beta decay*. Neutrinos have a very low probability of undergoing a nuclear reaction with other matter and therefore have a high probability of escaping from the Sun.

It was realized that if these neutrinos could be detected on Earth, they could be useful in determining the rate of fusion reactions in the Sun and also would help to answer other questions of fundamental physics. By devising very sensitive detectors and placing them deep underground, where they are shielded from other forms of radiation, it proved possible to detect these solar neutrinos. It came as a surprise that the neutrino flux detected was only about one-third of that expected from other estimates of the rate of fusion reactions in the Sun. It was predicted that there should be three types of neutrino: the *electron neutrino*, the *muon neutrino*, and the *tau neutrino*. The product of the proton–proton reaction in the Sun is an electron neutrino, but it gradually began to be suspected that the electron neutrinos might be changing into one of the other types on the way from the Sun to the Earth. The puzzle has recently been solved at a new laboratory built some 2000 m underground in a nickel mine in Canada. The Sudbury Neutrino Observatory (SNO) started operating in 1999 with a detector using 1000 tons of heavy water. These detectors are able to measure the total neutrino flux as well as the electron neutrino flux. It was shown that some of the electron neutrinos had indeed changed into the other types during their passage from the Sun to the Earth. When the total neutrino flux is measured, it is found to be in good agreement with the flux calculated from the Standard Solar model.

---

inside the sphere there has to be a balance between the outward pressure and the weight of the material above (outside) pressing downward (inward). The balance between compression due to gravity and outward pressure is called *hydrostatic equilibrium*. The same effect occurs in the Earth's atmosphere: The atmospheric pressure at sea level is due to the weight of the air above—this is the combined gravitational force acting on the air molecules. The atmosphere does not collapse to a very thin layer on the ground under the pull of gravity because the upward pressure of the compressed gas in the lower layers always balances the downward pressure of the upper layers.

In some ways the structure of the Sun is similar to that of the Earth, in the sense that it has a very dense core that contains most of the Sun's mass surrounded by less dense outer layers known as the *solar envelope* (Fig. 3.3). The temperature in

> ## BOX 3.2   The Carbon Cycle
>
> A second chain of nuclear reactions can convert hydrogen into helium. This was proposed independently by Carl von Weizacker and Hans Bethe in 1938, and it is now thought that it is the dominant process in stars that are hotter and more massive than the Sun. The reaction sequence is as follows:
>
> $$p + {}^{12}C \longrightarrow {}^{13}N \longrightarrow {}^{13}C + e^+ + \nu$$
> $$p + {}^{13}C \longrightarrow {}^{14}N$$
> $$p + {}^{14}N \longrightarrow {}^{15}O \longrightarrow {}^{15}N + e^+ + \nu$$
> $$p + {}^{15}N \longrightarrow {}^{12}C + {}^{4}He$$
>
> In the first stage a proton reacts with a ^{12}C nucleus to form nitrogen ^{13}N, which is unstable and decays to ^{13}C. Further stages build up through ^{14}N and ^{15}O to ^{15}N, which then reacts with a proton to form ^{12}C and ^{4}He. At the end of the sequence the ^{12}C has been recycled and can start another chain of reactions, so it acts as a catalyst. Overall four protons have been replaced with a single helium nucleus, so the energy release is the same as for the pp cycle.

the core is about 14 million degrees Celsius but falls quite rapidly within the radius—falling to about 8 million degrees Celsius at a quarter of the radius and to less than 4 million degrees Celsius at half the radius. Fusion reactions are very sensitive to temperature and density and take place only in the core. The fusion power density falls to 20% of its central value at 10% of the radius and to zero outside 20% of the radius.

Fusion energy is transported outward from the core as heat, first by radiation through the layers known as the *radiative zone*. But as the radiative zone cools with increasing distance from the core, it becomes more opaque and radiation becomes less efficient. Energy then begins to move by convection through huge cells of circulating gas several hundred kilometers in diameter in the *convective zone*. Finally the energy arrives at the zone that emits the sunlight that we see, the *photosphere*. This is a comparatively thin layer, only a few hundred kilometers thick, of low-pressure gases with a temperature of 6000°C. The composition, temperature, and pressure of the photosphere are revealed by the spectrum of sunlight. In fact, helium was discovered in 1896 by William Ramsey, who found features in the solar spectrum that did not belong to any gas known on Earth at that time. The newly discovered element was named helium in honor of Helios, the mythological Greek god of the Sun.

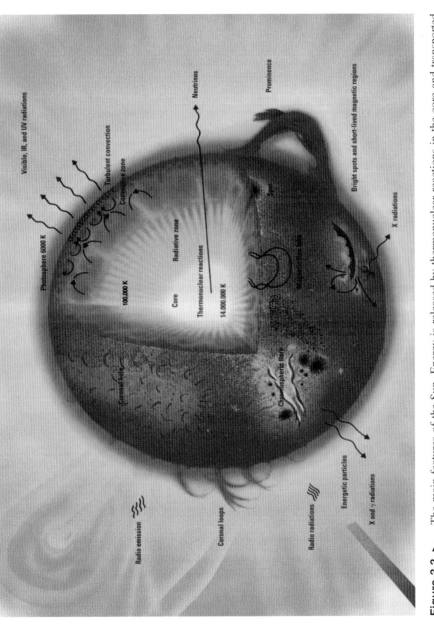

Figure 3.3 ▶ The main features of the Sun. Energy is released by thermonuclear reactions in the core and transported outward, first by radiation and then by convection, to the surface, from where it is radiated.

Gravity is a very weak force compared to the forces of nuclear physics, and it can confine a hot plasma only when the mass is very large. This is possible in the Sun and stars but not for the much smaller plasmas that we would like to confine on Earth. Also the fusion power density in the core of the Sun is very low, only 270 watts per cubic meter, compared to the megawatts per cubic meter required for a commercial power plant. Other methods of providing confinement have to be found, as will be discussed in later chapters.

3.4 The Formation of Heavier Atoms

In 1929, the American astronomer Edwin Hubble, by measuring the Doppler shift of the spectral lines from many stars and galaxies, discovered that the universe is expanding. He showed that the lines are shifted to the red end of the spectrum and hence that these bodies are moving away from the Earth. The effect is often known as the *red shift*. Hubble also showed that the further the objects are away, the faster they are moving. One explanation for this effect was that there was continuous creation of matter balancing the expansion, the *steady-state theory* of the universe. However, the presently accepted theory is that everything started about 13.7 billion years ago with a gigantic explosion known as the *Big Bang*. The idea of an expanding universe was proposed in 1927 by the Belgian cosmologist Georges Lemaître, and the model was described in detail by George Gamow (Fig. 3.4), Ralph Alpher, and Hans Bethe in 1948 in their famous "Alpher, Bethe, Gamow" paper. Their model predicted that there should be observable radiation left over from the Big Bang. This radiation, now known as the *Cosmic Microwave Background Radiation (CMBR)* (Box 3.3), was first observed by Arno Penzias and Robert Wilson in 1964 and found to be close to the predicted level. The predictions of the background radiation and of the correct abundances of hydrogen, helium, and lithium, which are now observed by spectroscopy of gas clouds and old stars, are the major successes of the Big Bang theory and are the justification for taking it to be the most likely explanation for the origin of the universe.

At the inconceivably high temperatures in the primeval fireball, mass and energy were continually interchanging. As the fireball expanded, it cooled rapidly, and at this stage the energy was converted permanently into mass — first as the underlying subnuclear building blocks were formed and then as these building blocks themselves combined to form protons and neutrons. Some deuterium and helium nuclei were formed, via the fusion reactions discussed earlier, when the conditions were suitable.

As the universe expanded it became too cold for these initial fusion reactions to continue, and the mix of different nuclei that had been produced was "frozen." It is calculated that this stage was reached only 4 minutes after the initial Big Bang. At this point the universe consisted of an expanding cloud composed mainly of hydrogen (75%) and helium (25%), with small amounts of deuterium and lithium.

Figure 3.4 ▶ George Gamow (1904–1968). Gamow was a Ukrainian, born in Odessa and educated in Leningrad, but in 1934 he immigrated to the US, where he was professor of physics, first at George Washington University and then at the University of Colorado. He was a prolific writer of books on science for the layperson, particularly on cosmology; many of these works are still in print.

BOX 3.3 Cosmic Microwave Background Radiation

The initial observations of microwave radiation, by Penzias and Wilson at Bell Telephone Laboratories, were made quite by accident. They were testing an antenna designed for communications satellites, and in order to check the zero level of their instrument they had pointed it at a region of the sky where they expected no radio sources. To their surprise they obtained a radiation signal that they could not explain. A few months later Jim Peebles at Princeton University heard of their results. He had been doing calculations based on the Big Bang theory, which predicted that the universe should be filled with a sea of radiation with a temperature less than 10 K. When they compared results, the observed radiation was in good agreement with predictions. As the measurements have improved, including more sophisticated instrumentation on satellites, it was found that not only was the intensity correctly predicted, but the measured CMBR also has precisely the profile of intensity versus frequency to be consistent with the Big Bang model.

In 1992 the COBE satellite showed for the first time that there are slight variations of the CMBR intensity with direction in the sky. These observations have been even further improved using the Wilkinson Microwave Anisotropy Probe in 2002. The radiation is calculated to have been generated 380,000 years after the Big Bang — over 13 billion years ago. It shows minute variations in the temperature of the CMBR. These tiny irregularities are the seeds of the cosmic structures that have been amplified by gravitational forces to become the stars and galaxies that we see today.

The universe today is known to contain 92 different elements ranging in mass from hydrogen to uranium. The theory of the Big Bang is quite explicit that nuclei much heavier than helium or lithium could not have been formed at the early stage. The obvious next question is, what is the origin of all the other elements, such as carbon, oxygen, silicon, and iron?

3.5 Stars and Supernovae

The formation of the stars occurs by gradual accretion, due to gravitational attraction in places where there were local density variations of the material spewed out from the Big Bang. As stars form they are compressed by gravity, and the interior heats until it becomes sufficiently hot for fusion reactions to start heating the star still further. There is an equilibrium where the internal pressure balances the compressive force due to gravity and, when all the fuel is burned up and the fusion reaction rate decreases, gravity causes the star to contract further. Stars form in a range of different sizes and this leads to a variety of different stellar life cycles. For a star of relatively modest size like our Sun, the life cycle is expected to be about 10 billion years. When all the hydrogen has been consumed and converted into helium, the Sun will cool down and shrink in size; with the Sun too cold for further fusion reactions, the cycle ends.

Larger stars heat to a higher temperature and therefore burn more rapidly. The fusion processes in these stars occur in a number of phases, forming more and more massive nuclei. After the first stage is completed and the hydrogen has been converted to helium, the bigger star's gravity is sufficiently strong that the star can be compressed further until the temperature rises to a value at which the helium nuclei start to fuse and form carbon in the core. This again releases energy, and the star gets even hotter. The mechanism by which helium burns was a puzzle for many years because the fusion of two helium nuclei would produce a nucleus of beryllium (^{8}Be), which is very unstable. It turns out that three helium nuclei have to join together to form a carbon nucleus (^{12}C), as explained in Box 3.4. When most of the helium has been consumed and if the star is big enough, further compression causes the temperature to rise again to the point at which the carbon burns, forming much heavier nuclei, such as neon (^{20}Ne) and magnesium (^{24}Mg). Neon is produced by the combination of two carbon nuclei followed by the release of a helium nucleus. In succession there are stages of neon burning and then silicon burning (Box 3.5). The reactions are shown schematically in Fig. 3.5.

The detailed verification of the models of the production of all the various elements has depended very largely on many years of study of the individual nuclear processes in physics laboratories. Rather as Aston's painstaking study of the precise masses of the elements led to the eventual realization of the source of nuclear energy, so the detailed measurements of the exact types and rates of nuclear reactions under a range of different conditions enabled the detailed evolution of the universe to become understood. Measurements that were initially made in the

BOX 3.4 | The Triple Alpha Process

When a star has converted most of the hydrogen into helium, the next stage would seem to be for two helium nuclei to combine. But this would produce ^{8}Be — a nucleus of beryllium with four protons and four neutrons that reverts back into two helium nuclei with a lifetime of less than 10^{-17} s. There is a further problem — even if the ^{8}Be nucleus is formed, the next stage in the chain,

$$^8\text{Be} + {}^4\text{He} \longrightarrow {}^{12}\text{C}$$

is not allowed because the energy cannot be removed as kinetic energy with a single reaction product without violating the law of conservation of momentum. Thus there appeared to be a bottleneck preventing the formation of the elements heavier than helium.

The English astronomer Fred Hoyle reasoned that nuclei heavier than He do in fact exist in nature — so there must be a way around the so-called *beryllium bottleneck*. He proposed that if the carbon nucleus has an excited state with energy of 7.65 MeV above the ground level of the carbon — exactly matching the energy released in the nuclear reaction — the reaction energy could be absorbed in the excited state, which could then decay to the ground state by the release of a gamma ray without any problems with momentum conservation. However no excited state was known at the time and so Hoyle approached William Fowler at the University of California at Berkeley and suggested to him that they conduct an experimental search for this state. Fowler agreed to look, and the excited state of carbon was found, thus verifying the mechanism by which the higher-mass nuclei are produced.

Overall the triple alpha process can be looked on as an equilibrium between three ^{4}He nuclei and the excited state of ^{12}C, with occasional leakage out into the ground state of ^{12}C. This is a very slow process and is viable only in a star with enormous quantities of helium and astronomical times scales to consider.

---

pursuit of fundamental academic research turned out to be crucially important in the understanding of the universe. Of course no one has made direct measurements inside the heart of a star; even the light that we can measure remotely with telescopes and analyze by spectroscopy comes from the star's surface.

At the end of a star's lifetime, when its nuclear fuel is exhausted, the release of fusion energy no longer supports it against the inward pull of gravity. The ultimate fate of a star depends on its size. Our Sun is a relatively small star and will end its life rather benignly as a *white dwarf*, as will most stars that begin life with mass up to about two to three times that of our Sun. If the star is more massive, its core will first collapse and then undergo a gigantic explosion known as a *supernova* and in so doing will release a huge amount of energy. This will cause a blast wave that ejects much of the star's material into interstellar space. Supernovae are relatively rarely observed, typically once in every 400 years in our own galaxy, but they can

BOX 3.5 Heavier Nuclei

After carbon has been produced, nuclei of higher mass can be formed by reactions with further alpha particles. Each of these nuclear reactions is less efficient in terms of energy production than the previous one, because the nuclei formed are gradually becoming more stable; see Figure 2.5. The temperature increases, the reactions proceed more quickly, and the time taken to burn the remaining fuel gets shorter. In a large star the time to burn the hydrogen might be 10 million years, while to burn the helium takes 1 million years, to burn the carbon only 600 years, and to burn the silicon less than one day! As each reaction dies down, due to the consumption of the fuel, gravity again dominates and the star is compressed. The compression continues until the star is sufficiently hot for the next reaction to start. It is necessary to reach successively higher temperatures to get the heavier elements to undergo fusion. The last fusion reactions are those that produce iron (^{56}Fe), cobalt, and nickel, the most stable of all the elements.

The principle reactions going on in various stages of the life of a massive star just prior to its explosive phase are shown in the following table (and schematically in Figure 3.5). The calculated density, temperature, and mass fraction in the various stages are shown, together with the composition in that stage.

Stage	Mass fraction	Temp (°C)	Density (kg·m^{-3})	Main reactions	Composition
I	0.6	1×10^7	10	^{1}H→ ^{4}He	^{1}H, ^{4}He
II	0.1	2×10^8	1×10^6	^{4}He→ ^{12}C, ^{16}O	^{4}He
III	0.05	5×10^8	6×10^6	^{12}C→ ^{20}Ne, ^{24}Mg	^{12}C, ^{16}O
IV	0.15	8×10^8	3×10^7	^{20}Ne→ ^{16}O, ^{24}Mg	^{16}O, ^{20}Ne, ^{24}Mg
V	0.02	3×10^9	2×10^9	^{16}O→ ^{28}Si	^{16}O, ^{24}Mg, ^{28}Si
VI	0.08	8×10^9	4×10^{12}	^{28}Si→ ^{56}Fe, ^{56}Ni	^{28}Si, ^{32}S

also be observed in other galaxies. Perhaps the most famous supernova historically is the one that was recorded by Chinese astronomers in 1054 AD. The remnants of this explosion are still observable and are known as the Crab Nebula. Since the Chinese observation, two similar explosions have taken place in our galaxy — one was observed by the famous Danish astronomer Tycho Brahe in 1572 and another by Johannes Kepler in 1604. Kepler, who was Brahe's pupil, discovered the laws of planetary motion. In 1987 the largest supernova to be observed since the invention of the telescope was seen in the Large Magellenic Cloud, the next nearest galaxy to the Milky Way. The last phase of the explosion occurred in a very short time in astronomical terms. It reached its brightest phase about 100 days after the original explosion and then started fading. Photographs taken of the original star before it exploded and then when the supernova was at its peak intensity are shown in Fig. 3.6.

Section 3.5 Stars and Supernovae

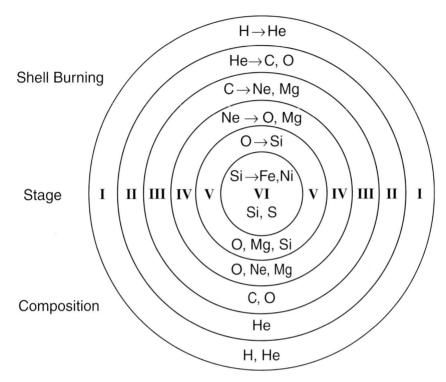

Figure 3.5 ▶ A schematic picture of the various fusion reactions occurring in the life of a large star, leading to the buildup of more and more massive nuclei. In the final stage the most stable elements around iron are formed. The normal chemical symbols for the elements are used. Just before the explosive stage, different reactions are occurring at different layers of the star, as shown.

The importance of supernovae in forming the elements is that their temperature is very high and large numbers of energetic neutrons are produced. These are ideal conditions for production of the higher-mass elements, from iron up to uranium. The energetic neutrons are absorbed by iron and nickel to form heavier atoms. All of the elements that have been created in the stars, both during the early burning phases and during the catastrophic phase of the supernovae, are redistributed throughout the galaxy by the explosion.

An idea of how an exploding supernova disperses in the universe is seen in the photograph of the Veil Nebula, Fig. 3.7. This is the remains of a supernova that exploded over 30,000 years ago. The material spreads out in fine wisps over an enormous volume. The dust created can then gather together to form new stars and planets *and us*. The second and subsequent generations of star systems, formed from the debris of supernovae, thus contain all the stable elements. The two paths by which primary and secondary stars can form are illustrated in Fig. 3.8. Our own solar system is an example of a secondary star system.

Figure 3.6 ▶ Photographs of the supernova that exploded in February 1987 in the Large Magellenic Cloud. This was the first supernova since the invention of the telescope that was bright enough to be visible to the naked eye. The view on the left shows the supernova at peak intensity, and the view on the right shows the same region before the star exploded.

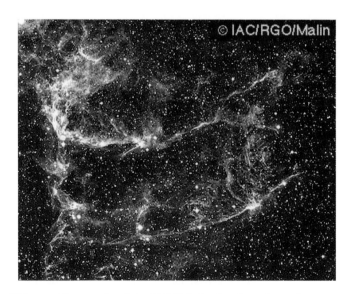

Figure 3.7 ▶ Photograph of the Veil Nebula showing wisps of matter remaining after the explosion of a supernova more than 30,000 years ago. The nebula is 15,000 light years away and has grown to enormous size, yet it maintains a fine filamentary structure.

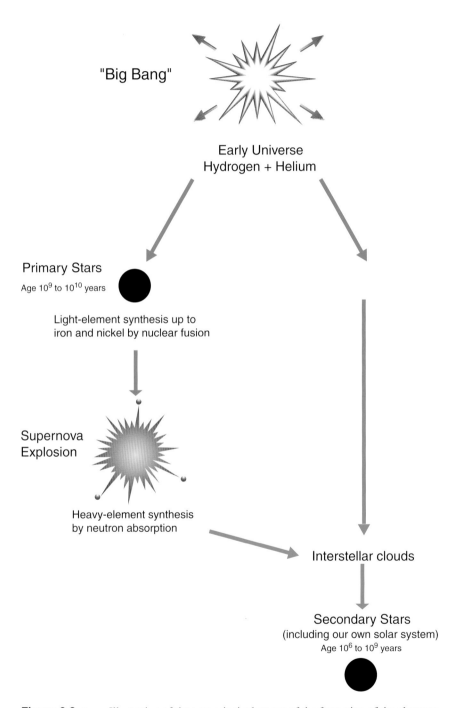

Figure 3.8 ▶ Illustration of the two principal stages of the formation of the elements showing how the first-generation stars contained only the light elements from the Big Bang, while second-generation stars contain the heavier elements from the supernovae explosions.

The remains of the cores of supernovae are thought to form strange objects known as *neutron stars* and *black holes*. The pull of gravity becomes so enormous that matter is squeezed tightly together, and nuclei, protons, and electrons all convert into neutrons. A neutron star has roughly one and a half times the mass of our Sun crammed in a ball about 10 kilometers in radius. Its density is therefore 100 trillion times the density of water; at that density, all the people on Earth would fit into a teaspoon! As the core gets smaller, it rotates faster and faster, like a skater who pulls his or her arms in. Strong radio radiation is emitted and can be detected on Earth as pulses, and so neutron stars are also known as *pulsars*. Neutron stars are relatively rare, only about one in a thousand stars, and the nearest one is probably at least 40 million light years away.

The stars that eventually become neutron stars are thought to start out with about 15–30 times the mass of our Sun. Stars with even higher initial masses are thought to become black holes — a region of space in which the matter forming it is crushed out of existence. The mass of a black hole is so large and the resulting gravitational field at its surface so strong that nothing, not even light, can escape.

The roles of nuclear fusion in the universe can therefore be summarized under two main headings. Firstly, fusion is the source of all the energy in the stars, thus supplying the energy by which we on Earth, and possibly other civilizations on the planets of other stars, survive. Secondly, fusion is responsible for the formation of the elements out of the primeval hydrogen. Some of the light elements (mainly hydrogen and helium) were formed by fusion in the Big Bang at the very start of the universe. The elements in the lower half of the periodic table are formed by the steady burning of the largest and hottest stars, and the very heaviest of the elements are produced by the very brief but intense reactions in the exploding supernovae.

Chapter 4

Man-Made Fusion

4.1 Down to Earth

Chapter 3 discussed the processes by which the Sun and stars release energy from fusion. There is no doubt that fusion works, but the obvious question is, can fusion energy be made useful to mankind? The physics community had been skeptical at first about the possibility of exploiting nuclear energy on Earth; even Rutherford had gone so far as to call it "moonshine." However, speculation on the subject abounded from the days when it was suspected that nuclear processes might be important for the stars.

Impetus was added when the first atom bombs were exploded in the closing stages of World War II, with the dramatic demonstration that nuclear energy could indeed be released. If nuclear fission could release energy, why not nuclear fusion? The present chapter discusses the basic principles of how fusion energy might be exploited on Earth.

The chain of reactions in the Sun starts with the fusion of two protons — the nuclei of the common form of hydrogen — to form a nucleus of *deuterium* — the heavier form of hydrogen. When two protons fuse, one of them has to be converted into a neutron. This is the most difficult stage in the chain of reactions that power the Sun, and it takes place much too slowly to be a viable source of energy on Earth. However, after the slow first step, the fusion reactions only involve rearranging the numbers of protons and neutrons in the nucleus, and they take place much more quickly. So things look more promising if one starts with deuterium. Though deuterium is rare in the Sun, where it is burned up as fast as it is produced, on Earth there are large amounts of this form of hydrogen remaining from earlier cosmological processes. About one in every 7000 atoms of hydrogen is deuterium, and these two isotopes can be separated quite easily. The Earth has a very large amount of hydrogen, mainly as water in the oceans, so although deuterium is rather dilute, the total amount is virtually inexhaustible (Box 4.1).

33

> **BOX 4.1** **Source of Deuterium**
>
> The reaction between deuterium and tritium has the fastest reaction rate and requires the lowest temperature of all the fusion reactions, so it is the first choice for a fusion power plant. Adequate sources of deuterium and tritium are thus important, independent of what type of confinement system, magnetic or inertial, is employed.
>
> One gram of deuterium will produce 300 GJ of electricity, and providing for all of the world's present-day energy consumption (equivalent to about 3×10^{11} GJ per year) would require about 1000 tons of deuterium a year. The source of deuterium is straightforward because about 1 part in 6700 of water is deuterium, and 1 gallon of water used as a source of fusion fuel could produce as much energy as 300 gallons of gasoline. When all the water in the oceans is considered, this amounts to over 10^{15} tons of deuterium — enough to supply our energy requirements indefinitely. Extracting deuterium from water is straightforward using electrolysis (see Box 8.1), and the cost of the fuel would be negligible compared to the other costs of making electricity.

The fusion reaction between two deuterium nuclei brings together two protons and two neutrons that can rearrange themselves in two alternative ways. One rearrangement produces a nucleus that has two protons and a single neutron. This is the rare form of helium known as *helium-3* (see Fig. 2.2). There is a neutron left over. The alternative rearrangement produces a nucleus with one proton and two neutrons. This is the form of hydrogen known as *tritium*, which has roughly three times the mass of ordinary hydrogen. In this case a proton is left over. These reactions are shown schematically in Fig. 4.1. Energy is released because the sum of the masses of the rearranged nuclei is slightly smaller than the mass of two deuterium nuclei, as in the fusion reactions in the Sun.

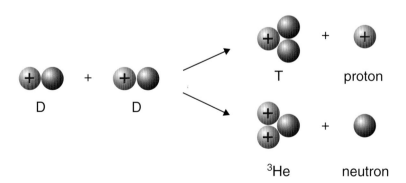

Figure 4.1 ▶ The two alternative branches of the fusion of two deuterium nuclei.

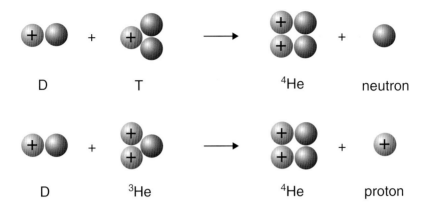

Figure 4.2 ▶ The reactions between deuterium and tritium or helium-3, forming helium-4.

The tritium and the helium-3 produced in these reactions can also fuse with deuterium. This time there are five nuclear particles to rearrange — two protons and three neutrons in the case of the reaction between deuterium and tritium or three protons and two neutrons in the case of deuterium plus helium-3. The result in both cases is a nucleus with two protons and two neutrons. This is the common form of helium with four units of mass — *helium-4*. It is an inert gas that can be used to fill balloons and airships. There is either a free neutron or a free proton left over. These reactions are shown schematically in Fig. 4.2.

All of these reactions are used in experiments to study fusion. The reaction between deuterium and tritium, usually abbreviated as DT, requires the lowest temperature to get it started and therefore is considered to be the best candidate for a fusion power plant. Tritium does not occur naturally on Earth because it is radioactive, decaying with a half-life of 12.3 years. Starting with a fixed quantity of tritium today, only half of it will remain in 12.3 years time, there will be only a quarter after 24.6 years, and so on. Tritium will have to be manufactured as a fuel. In principle this can be done by allowing the neutron that is produced in the DT reaction to react with the element lithium (see Box 4.2). Lithium has three protons in its nucleus and exists in two forms — one with three neutrons, known as *lithium-6*, and one with four neutrons, known as *lithium-7*. Both forms interact with neutrons to produce tritium and helium. In the first reaction energy is released, but energy has to be put into the second reaction. The basic fuels for a fusion power plant burning deuterium and tritium thus will be ordinary water and lithium. Deuterium will be extracted from water, and tritium will be produced from lithium. Both basic fuels are relatively cheap, abundant, and easily accessible. The waste product will be the inert gas helium. The overall reaction is shown schematically in Fig. 4.3. The economics of fusion will be discussed in more detail in Chapter 12.

> **BOX 4.2** **Tritium Breeding Reactions**
>
> The most convenient way to make tritium is in the reaction between neutrons and lithium. There are two possible reactions, one with each of the naturally occurring isotopes, ^{6}Li and ^{7}Li:
>
> $$^6\text{Li} + n \rightarrow {}^4\text{He} + T + 4.8 \text{ MeV}$$
>
> $$^7\text{Li} + n \rightarrow {}^4\text{He} + T + n - 2.5 \text{ MeV}.$$
>
> The ^{6}Li reaction is most probable with a slow neutron; it is exothermic, releasing 4.8 MeV of energy. The ^{7}Li reaction is an endothermic reaction, only occurring with a fast neutron and absorbing 2.5 MeV of energy. Natural lithium is composed 92.6% of ^{7}Li and 6.4% of ^{6}Li. A kilogram of lithium will produce 1×10^5 GJ of electricity.

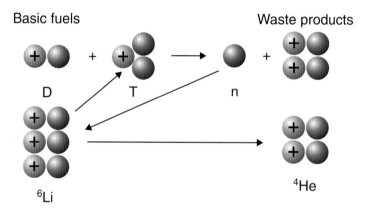

Figure 4.3 ▶ The overall fusion reaction. The basic fuels are deuterium and lithium; the waste product is helium.

4.2 Getting It Together

In order to initiate these fusion reactions, two nuclei have to be brought very close together, to distances comparable to their size. Nuclei contain protons and so they are positively charged. Charges of the same polarity, in this case two positive charges, repel each other; there is thus a strong electric force trying to keep the two nuclei apart. Only when the two nuclei are very close together does an attractive nuclear force become strong enough to counter the electric force that is trying to keep them apart. This effect is shown schematically in Fig. 4.4 — which plots

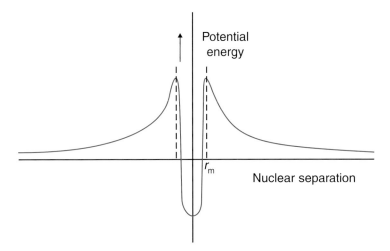

Figure 4.4 ▶ A diagram illustrating the potential energy of two nuclei as their distance apart is varied. When far apart they repel each other — their electrostatic repulsion increases as they get closer together. When they become very close, a nuclear attraction becomes effective and the potential energy drops.

potential energy against the distance separating the two nuclei. Potential energy is the form of energy that a ball has on top of a hill. Bringing the two nuclei together is rather like trying to get a golf ball to roll up a hill and fall into a hole on the top. The ball has to have enough energy to climb the hill before it can fall into the hole. In this case the hill is very much steeper and the hole much deeper and smaller than anything one would find on a golf course. Fortunately physics comes to our aid. The laws of quantum mechanics that determine how nuclei behave at these very small distances can allow the "ball" to tunnel partway through the hill rather than having to go all the way over the top. This makes it a bit easier, but even so a lot of energy is needed to bring about an encounter close enough for fusion.

Physicists measure the energy of atomic particles in terms of the voltage through which they have to be accelerated to reach that energy. To bring about a fusion reaction requires acceleration by about 100,000 volts. The probability that a fusion reaction will take place is given in the form of a *cross section*. This is simply a measure of the size of the hole into which the ball has to be aimed. The cross sections for the three most probable fusion reactions are shown in Fig. 4.5. In fusion golf, the effective size of the hole depends on the energy of the colliding nuclei. For DT, the cross section is largest when the nuclei have been accelerated by about 100,000 volts (to energy of 100 keV); it decreases again at higher energies. Figure 4.5 shows why the DT reaction is the most favorable — it offers the highest probability of fusion (the largest cross section) at the lowest energy. Even then the hole is very small — with an area of about 10^{-28} square meters.

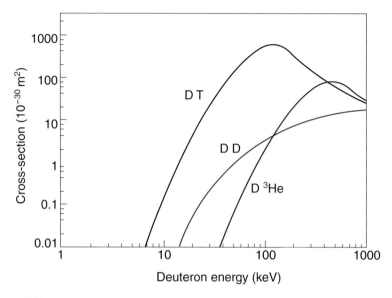

Figure 4.5 ▶ The probability that a fusion reaction will take place (cross section) for a range of energies of deuterium ions. The data for three reactions are shown: deuterium plus deuterium, deuterium plus tritium, and deuterium plus helium-3. At lower energies the probability for the DT reaction is much higher than for the other two reactions.

Voltages of hundreds of thousands of volts sound rather high compared to the hundreds of volts of a normal domestic electricity supply. However, in 1930 John Cockroft and Ernest Walton, working in Rutherford's laboratory at Cambridge University, designed and built a particle accelerator capable of generating these voltages. This sort of equipment is now commonplace in physics laboratories. In fact, physicists studying the structures within protons and neutrons use accelerators that take particles into the *gigavolt* range of energies, that is, thousands of millions of volts, and they have plans to build *teravolt* machines, that is, millions of millions of volts.

Accelerating nuclei to the energies needed for fusion is not difficult in the laboratory. It is relatively easy to study fusion reactions by bombarding with accelerated deuterium nuclei a solid target containing tritium. This is how the cross sections shown earlier were measured by Marcus Oliphant and Paul Hartek in Cambridge, UK, in 1934. The problem lies with the very small cross section of the "fusion hole" and the very steep "hill." Most of the accelerated nuclei bounce off the "hill" and never get close enough to the target nucleus to fuse. The energy that has been invested in accelerating them is lost. Only a tiny fraction of collisions (1 in 100 million) actually results in a fusion event. To return to the golfing analogy, it is rather like firing ball after ball at the hill in the hope that one will be lucky enough to find its way over the top and into the hole. Very few will make it when the par for the hole is 100 million; most balls will roll off the hill and be lost. In the case of fusion the problem is not so much the number of lost balls but the amount of energy that is lost with them.

A better way has to be found. Clearly what is needed is a way to collect all the balls that roll off the hill and, without letting them lose energy, to send them back again and again up the slope until they finally make it into the hole. Leaving the golf course and moving indoors to the billiard or pool table illustrates how this might be done. If a ball is struck hard, it can bounce back and forth around the pool table without loosing energy (assume that the balls and the table are frictionless). Doing this with a large number of balls in motion at the same time will allow balls to scatter off each other repeatedly without losing energy. Occasionally two balls will have the correct energies and be moving on exactly the right trajectories to allow them to fuse together when they collide. It is important to remember however, that this happens only once in every 100 million encounters.

This picture of balls moving about randomly and colliding with each other is rather like the behavior of a gas. The gas particles — they are usually *molecules* — bounce about quite randomly, off each other and off the walls of the container, without losing any overall energy. Individual particles continually exchange energy with each other when they collide. In this way, there will always be some particles with high energies and some with low energies, but the average energy stays constant. The *temperature* of the gas is a measure of this average energy.

These considerations suggest a better way to approach fusion; take a mixture of deuterium and tritium gas and heat it to the required temperature. Known as *thermonuclear* fusion, this is to be clearly distinguished from the case where individual nuclei are accelerated and collided with each other or with a stationary target. A temperature of about 200 million degrees Celsius is necessary to give energies high enough for fusion to occur in a sufficiently high fraction of the nuclei. It is difficult to get a feel for the magnitude of such high temperatures. Remember that ice melts at 0° Celsius, water boils at a 100°C, iron melts at around 1000°C, and everything has vaporized at 3000°C. The temperature of the core of the Sun is about 14 million degrees. For fusion reactions, it is necessary to talk in terms of hundreds of millions of degrees. To put 200 million degrees on a familiar scale would require an ordinary household thermometer about 400 kilometers long!

Collisions in the hot gas quickly knock the electrons off the atoms and produce a mixture of nuclei and electrons. The gas is said to be *ionized*, and it has a special name — it is called a *plasma* (Fig. 4.6). Plasma is the fourth state of matter — solids melt to form liquids, liquids evaporate to form gases, and gases can be ionized to form plasmas. Gases exist in this ionized state in many everyday conditions, such as in fluorescent lamps and even in open flames, although in these examples only a small fraction of the gas is ionized. In interstellar space practically all matter is in the form of fully ionized plasma, although the density of particles is generally very low. Since plasmas are a fundamental state of matter, their study is justified as pure science on the same basis as research into the solid, liquid, or gaseous state. At the high temperatures required for fusion the plasma is

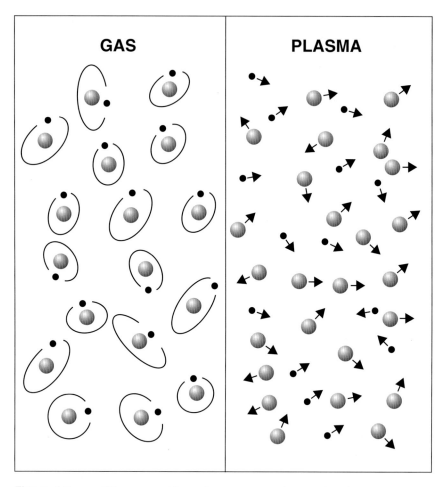

Figure 4.6 ▶ When a gas (shown here as an atomic gas, though many gases are molecular) is heated to high temperature it breaks up into a mixture of negatively charged electrons and positively charged nuclei or ions.

fully ionized and consists of a mixture of negative electrons and positive nuclei. Equal numbers of negative and positive charge must be present; otherwise the unbalanced electric forces would cause the plasma to expand rapidly. The positive nuclei are called *ions*, and this term will be used from now on. One important property of plasmas is that, with all these electrically charged particles, they can conduct electricity. The electrical conductivity of hydrogen plasma, at the temperatures required for fusion to occur, is about 10 times higher than that of copper at normal temperature.

What about the walls? The temperature of the ions has to be so high that there is no possibility of containing the hot plasma in any conventional vessel.

Even the most refractory materials, such as graphite, ceramics, or tungsten, would evaporate. There are two options. One is to use a magnetic field to form a barrier between the hot fuel and the wall. The electrical charges on ions and electrons prevent them from moving directly across a magnetic field. When their motion tries to take them across the field, they simply move around in circles. They can move freely along the direction of the field, and so the overall motion is a spiral line (a helix) along the direction of the field. In this way a magnetic field can be used to guide the charged particles and prevent them from hitting the surrounding solid walls. This is called *magnetic confinement*. The second option is to compress the fusion fuel and heat it so quickly that fusion takes place before the fuel can expand and touch the walls. This is called *inertial confinement*. It is the principle used in the hydrogen bomb and in attempts to produce fusion energy using lasers. These two options will be described in more detail in the following chapters.

4.3 Breaking Even

One of the fundamental questions is to determine the conditions required for a net energy output from fusion. Energy is needed to heat the fuel up to the temperature required for fusion reactions, and the hot plasma loses energy in various ways. Clearly there would be little interest in a fusion power plant that produces less energy than it needs to operate. John Lawson (Fig. 4.7), a physicist at the UK Atomic Energy Establishment at Harwell, showed in the early 1950s that "it is necessary to maintain the plasma density multiplied by the confinement time greater than a specified value." The plasma density (usually denoted by n) is the number of fuel ions per cubic meter. The *energy confinement time*, usually denoted by the Greek letter tau (τ_E), is more subtle. It is a measure of the rate at which energy is lost from the plasma and is defined as the total amount of energy in the plasma divided by the rate at which energy is lost. It is analogous to the time constant of a house cooling down when the central heating is switched off. Of course the plasma is not allowed to cool down; the objective is to keep it at a uniformly high temperature. Then the energy confinement time is a measure of the quality of the magnetic confinement. Just as the house cools down more slowly when it is well insulated, so the energy confinement time of fusion plasma is improved by good magnetic "insulation." Lawson assumed that all the fusion power was taken out as heat and converted to electricity with a specified efficiency (he took this to be about 33% which is a typical value for a power plant). This electricity would then be used to heat the plasma. Nowadays the calculation for magnetic-confinement fusion makes slightly different assumptions but arrives at a similar conclusion.

The DT reaction produces a helium nucleus—usually known as an *alpha particle*—and a neutron. The energy released by the fusion reaction is shared between the alpha particle, with 20% of the total energy, and the neutron, with 80%. The neutron has no electric charge, and so it is not affected by the magnetic

Figure 4.7 ▶ John Lawson (b. 1923) explaining energy balance requirements at a meeting of the British Association in Dublin in 1957. Experimental work was still secret at this time, and the meeting was the first time that energy balance had been publicly discussed. Lawson worked in the nuclear fusion program from 1951 to 1962, but most of his career was devoted to the physics of high-energy accelerators.

field. It escapes from the plasma and slows down in a surrounding structure, where it transfers its energy and reacts with lithium to produce tritium fuel. The fusion energy will be converted into heat and then into electricity. This is the output of the power plant. The alpha particle has a positive charge and is trapped by the magnetic field. The energy of the alpha particle can be used to heat the plasma. Initially an external source of energy is needed to raise the plasma temperature. As the temperature rises, the fusion reaction rate increases and the alpha particles provide more and more of the required heating power. Eventually the alpha heating is sufficient by itself and the fusion reaction becomes self-sustaining. This point is called *ignition*. It is exactly analogous to using a gas torch to light a coal fire or a barbecue. The gas torch provides the external heat until the coal is at a high enough temperature that the combustion becomes self-sustaining.

The condition for ignition in magnetic confinement is calculated by setting the alpha particle heating equal to the rate at which energy is lost from the plasma. This is slightly more stringent than the earlier version proposed by Lawson because only 20% of the fusion energy (rather than 33%) is used to heat the plasma. The ignition condition has the same form as the Lawson criterion, and the two are frequently confused. The product of density and confinement time must be larger than some specified value, which depends on the plasma temperature and has a minimum value (see Fig. 4.8) in DT at about 30 keV (roughly 300 million degrees). Written in the units *particles per cubic meter × seconds*, the condition

Section 4.3 Breaking Even

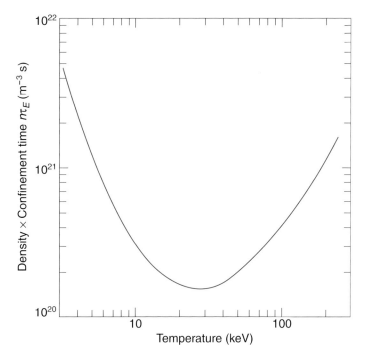

Figure 4.8 ▶ The ignition criterion: the value of the product of density and confinement time $n\tau_E$, necessary to obtain plasma ignition, plotted as a function of plasma temperature T. The curve has a minimum at about 30 keV (roughly 300 million °C).

for ignition is

$$n \times \tau_E > 1.7 \times 10^{20} \text{ m}^{-3} \text{ s}$$

However, due to the way that the fusion cross sections and other parameters depend on temperature, it turns out that the best route to ignition is at slightly lower temperatures. In the range 10–20 keV (100 million to 200 million degrees), the ignition condition can be written in a slightly different form that includes the temperature (see Box 4.3),

$$nT\tau_E > 3 \times 10^{21} \text{ m}^{-3} \text{ keV s}$$

The units are *particles per cubic meter × kilo-electron volts × seconds*. This can be expressed in different units that are a bit more meaningful for a nonspecialist. The product of density and temperature is the pressure of the plasma. The ignition condition then becomes: *plasma pressure (P) × energy confinement time (τ_E) must be greater than 5.*

BOX 4.3 Conditions for Confinement

The conditions for DT magnetic-confinement fusion to reach ignition and run continuously are calculated by setting the alpha particle heating equal to the rate at which energy is lost from the plasma. Each alpha particle transfers 3.5 MeV to the plasma, and the heating power per unit volume of the plasma (in MWm^{-3}) is $P_\alpha = n_D n_T \overline{\sigma v} \, k \, 3.5 \times 10^3$. The DT fusion reaction rate $\overline{\sigma v}$ (m^3 s^{-1}) is the cross section σ averaged over the relative velocities v of the colliding nuclei at temperature T (keV), and n_D and n_T are the densities (m^{-3}) of D and T fuel ions. The reaction is optimum with a 50:50 fuel mixture, so $n_D = n_T = \frac{1}{2}n$, where n is the average plasma density and

$$P_\alpha = \frac{1}{4} n^2 \overline{\sigma v} \, k \, 3.5 \times 10^3 \text{ MWm}^{-3}$$

The loss from the plasma is determined as follows. The average energy of a plasma particle (ion or electron) at temperature T is $(3/2)kT$ (corresponding to $\frac{1}{2}kT$ per degree of freedom). There are equal numbers of ions and electrons, so the total plasma energy per unit volume is $3nkT$. Here k is Boltzmann's constant, and when we express T in keV it is convenient to write $k = 1.6 \times 10^{-16}$ J/keV. The rate of energy loss from the plasma P_L is characterized by an energy-confinement time τ_E such that $P_L = 3nkT/\tau_E$. Setting the alpha particle heating equal to the plasma loss gives

$$n\tau_E = (12/3.5 \times 10^3)(T/\overline{\sigma v}) \text{ m}^{-3} \text{ s}$$

The right-hand side of this equation is a function only of temperature and has a minimum around $T = 30$ keV, where

$$(T/\overline{\sigma v}) \approx 5 \times 10^{22} \text{ keV m}^{-3} \text{ s,}$$

so the required value of $n\tau_E$ would be

$$n\tau_E \approx 1.7 \times 10^{20} \text{ m}^{-3} \text{ s}$$

(Fig. 4.8).

Usually τ_E is also a function of temperature (see Box 10.4), and the optimum temperature comes somewhat lower than 30 keV. Fortunately we can take advantage of a quirk of nature. In the temperature range 10–20 keV the DT reaction rate $\overline{\sigma v}$ is proportional to T^2. Multiplying both sides of the equation for $n\tau_E$ by T makes the right-hand side ($T^2/\overline{\sigma v}$) independent of temperature, while the left-hand side becomes the triple product $nT\tau_E$:

$$nT\tau_E = \text{const} \approx 3 \times 10^{21} \text{ m}^{-3} \text{ keV s}^{-1}$$

The precise value in fact depends on the profiles of plasma density and temperature and on other issues, like the plasma purity. A typical value

Section 4.3 **Breaking Even**

> **BOX 4.3** (continued)
>
> taking these factors into account would be
>
> $$nT\tau_E \approx 6 \times 10^{21} \text{ m}^{-3} \text{ keV s}^{-1}$$
>
> It is important to stress that the triple product is a valid concept only for T in the range 10–20 keV.
>
> The conditions required for a pulsed system (as in inertial-confinement fusion) can be expressed in a similar form if τ_E is defined as the pulse duration and the steady-state balance between alpha particle heating and energy loss is replaced by the assumption that all of the fusion energy is extracted after each pulse, converted into electricity and some of the output has to be used to heat the fuel for the next pulse. The efficiency of the conversion and heating cycles for inertial confinement is discussed further in Boxes 7.1, 7.2, and 11.3.

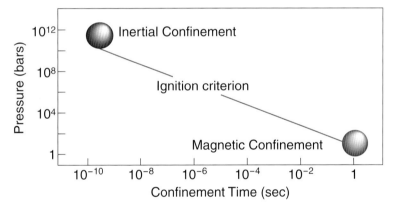

Figure 4.9 ► The conditions required for fusion plotted in terms of plasma pressure (in bars) against confinement time (in seconds). The regions for inertial-confinement and magnetic-confinement fusion are shown. In both cases a temperature in the range 10–20 keV (roughly 100–200 million degrees Celsius) is required.

The units are now *bars* × *seconds*, and one bar is close to the Earth's atmospheric pressure. The relationship between plasma pressure and confinement time is shown in Fig. 4.9.

For magnetic-confinement fusion, an energy-confinement time of about 5 seconds and a plasma pressure of about 1 bar is one combination that could meet this condition. It is at first sight surprising to find that the pressure of very hot plasma is close to that of the atmosphere. However, pressure is density multiplied by temperature, so a relatively low density balances the very high temperature of fusion plasma. For inertial-confinement fusion, the time is much shorter, typically less than 1 billionth (10^{-9}) of a second, and the pressure has to be correspondingly higher — more than 5 billion times atmospheric pressure. The fusion fuel has to be compressed until the density is about 1000 times higher than water.

The triple product known as $n\ T\ tau$ is used as the figure of merit against which the results of fusion experiments are compared (but note that this is valid only for temperatures in the range 10–20 keV). Progress toward ignition has required a long and difficult struggle, but now the goal is well within sight. Temperatures higher than 30 keV (300 million degrees) have been reached in some experiments, and confinement times and densities are in the right range. The most recent results in magnetic confinement, which will be described in later chapters, have pushed the triple product $nT\tau$ up to a value that is only a factor of 5 short of ignition. The best results in inertial confinement are about a factor of 10 lower than magnetic confinement.

▶ Chapter 5

Magnetic Confinement

5.1 The First Experiments

As with most developments in science, many people were thinking along similar lines at the same time, so it is hard to give credit to any single person as the "discoverer" of magnetic-confinement fusion. There is some evidence of speculative discussions of the possibility of generating energy from the fusion reactions before and during World War II. Certainly there had been discussions about the basic principles of fusion among the physicists building the atom bomb in Los Alamos. They had more urgent and pressing priorities at the time, and for various reasons they did not pursue their embryonic ideas on fusion immediately after the end of the war. However, many of these scientists did return later to work on fusion.

The first tangible steps were taken in the UK. In 1946, George Thomson (Fig. 5.1) and Moses Blackman at Imperial College in London registered a patent for a thermonuclear power plant. Their patent was quickly classified as secret, so the details were not made public at the time. In essence, the patent outlined a plan for a hot plasma confined by a magnetic field in a doughnut-shaped vessel that superficially looks remarkably like present-day fusion experiments. The proper geometric term for this doughnut shape — like an inflated automobile tire — is a *torus*. With the benefit of present knowledge of the subject, it is clear that this early idea would not have worked — but it displays remarkable insight for its day. This proposal provoked much discussion and led to the start of experimental fusion research at Imperial College.

A parallel initiative had been started in 1946 in the Clarendon Laboratory at Oxford University. Peter Thonemann (Fig. 5.1) had come to Oxford from Sydney University in Australia, where, earlier in the century, the so-called *pinch effect* had been discovered. The heavy electric current that had flowed through a hollow lightning conductor during a storm had been found to have permanently squashed it.

Figure 5.1 ▶ George Thomson (on the left) and Peter Thonemann at a conference in 1979. Thomson was awarded the Nobel Prize in Physics in 1937 for demonstrating the wave characteristics of electrons.

If the pinch effect was strong enough to compress metal, perhaps it could confine plasma. The basic idea is quite simple and is shown schematically in Fig. 5.2. When an electric current flows through a conductor — in this case the plasma — it generates a magnetic field that encircles the direction of the current. If the current is sufficiently large, the magnetic force will be strong enough to constrict, or *pinch*, the plasma and pull it away from the walls. In a straight tube the plasma will rapidly escape out of the open ends. However, if the tube is bent into a torus, it is possible in principle to create a self-constricted plasma isolated from contact with material surfaces. This is discussed in more detail in Box 5.1.

Thonemann worked with a series of small glass tori. The air inside the glass torus was pumped out and replaced with hydrogen gas at a much lower pressure than the atmosphere. This gas could be ionized to make plasma. A coil of wire wrapped around the outside of the torus was connected to a powerful radio transmitter. The current flowing through the external coil induced a current to flow in the plasma inside the torus. After a few years this system was replaced by a more efficient arrangement using an iron transformer core and a high-voltage capacitor. When the capacitor was discharged through the primary coil of the transformer, it induced a current in the plasma, which formed the secondary coil.

Section 5.1 **The First Experiments**

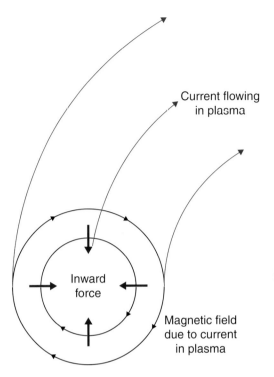

Figure 5.2 ▶ Schematic drawing showing the physical mechanism that causes an electric current to compress the conductor through which it is flowing. The current in the plasma flows round the torus, producing the magnetic field. The force due to the interaction of the current and its own field is directed inward.

| BOX 5.1 | **Magnetic Confinement** |

A charged particle in a uniform magnetic field moves freely in the direction parallel to the field, but there is a force in the transverse direction that forces the particle into a circular orbit. The combined motion of the particle is a spiral, or helical, path along the direction of the magnetic field; see Figure 5.3a. The transverse radius of the helical orbit is known as the *Larmor radius* ρ_e (sometimes also called the *gyro radius* or *cyclotron radius*), and it depends on the charge, mass, and velocity of the particle as well as the strength of the magnetic field. The Larmor radius of an electron

$$\rho_e = 1.07 \times 10^{-4} T_e^{0.5}/B$$

where T_e is the temperature, in kiloelectron-volts, and B is the magnetic field, in teslas. An ion with charge number Z and mass number A has a Larmor radius

$$\rho_i = 4.57 \times 10^{-3} (A^{0.5}/Z) T_i^{0.5}/B$$

> **BOX 5.1** (continued)

Thus a deuterium ion orbit is about 60 times larger than the orbit of an electron at the same temperature and magnetic field. (See Fig. 5.3).

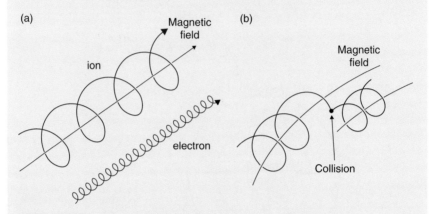

Figure 5.3 ▶ (a) Schematic of an ion and an electron gyrating in a straight magnetic field; (b) an ion collision resulting in the ion being displaced to a new orbit.

As in a gas mixture, where the total pressure is the sum of the partial pressures of the constituents, a hot plasma exerts an outward pressure that is the sum of the kinetic pressures of the electrons and the ions; thus $P = n_e k T_e + n_i k T_i$, where $k = 1.38 \times 10^{-23}$ J/°K, or 1.6×10^{-16} J/keV, is Boltzmann's constant. For simplicity we can take $n_e = n_i$ and $T_e = T_i$, but this is not always true. In magnetic confinement, the outward pressure of the plasma has to be balanced by an inward force — and it is convenient to think of the magnetic field exerting a pressure equal to $B^2/2\mu_0$, where B is the magnetic field strength, in teslas, and $\mu_0 = 4\pi \times 10^{-7}$ H/m is the permeability of free space. The ratio of plasma pressure to magnetic pressure is defined by the parameter $\beta = 2\mu_0 P/B^2$. There have been many attempts to develop magnetic-confinement configurations with $\beta \approx 1$, but the most successful routes to fusion, tokamaks and stellarators, require for stability rather low values of β, typically only a few percent.

In an ideal magnetic-confinement system, charged particles can cross the magnetic field only as a result of collisions with other particles. Collisions cause particles to be displaced from their original orbit onto new orbits (Fig. 5.3b), and the characteristic radial step length is of the order of the Larmor radius. Collisions cause an individual particle to move randomly either inward or outward, but when there is a gradient in the particle density there is a net outward diffusion of particles. The diffusion coefficient has the form ρ^2/t_c, where t_c is the characteristic time between collisions. Ions, with

Section 5.1 **The First Experiments**   51

BOX 5.1 (continued)

an orbit radius significantly larger than the electrons, would be expected to diffuse much faster than electrons, but they are prevented from doing so by the requirement that the plasma should remain neutral. A radial electric field is set up that impedes the ion cross-field diffusion rate so that it matches that of the electrons—this is known as the *ambipolar* effect. This simple classical picture fails, however, to account for the losses actually observed in magnetically-confined plasmas, and we need to look to other effects (Box 10.3).

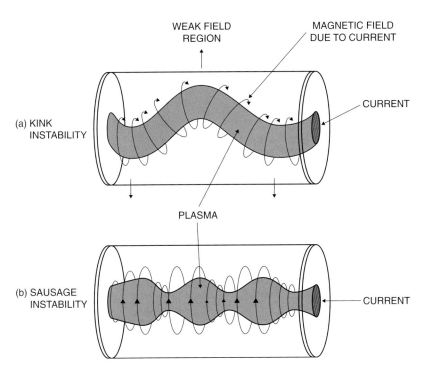

Figure 5.4 ▶ An illustration of some of the ways in which a plasma "wriggles" when an attempt is made to confine it in a magnetic field. If the plasma deforms a little, the outer side of the field is stretched and weakened. This leads to growth of the deformation and hence to instability.

Of course it turned out that the creation of a plasma was not so simple in practice, and it was soon found that the plasma was dreadfully unstable. It wriggled about like a snake and quickly came into contact with the torus walls, as shown in Fig. 5.4. Some, but not all, of these instabilities were tamed by adding another magnetic field from additional coils wound around the torus. It was hard to measure the temperatures. Estimates showed that, though high by everyday standards, they fell far short of the hundreds of millions of degrees that were required for fusion.

5.2 Behind Closed Doors

Fusion research looked promising, not only to the scientists but also to governments. As well as producing energy, one application that seemed attractive at the time was that fusion would generate large numbers of neutrons. It was thought that these could be used to make the plutonium needed for nuclear weapons more efficiently and quickly than it could be produced in fission reactors. So fusion research was soon classified as secret and moved away from universities into more secure government research centers, like that at Harwell near Oxford, England.

A curtain of secrecy came down, and little more was heard about fusion over the next few years. A slightly bizarre event occurred in 1951 when Argentine President Peron announced that an Austrian physicist working in Argentina had made a breakthrough in fusion research. The details were never revealed, and the claim was later found to be false. But the publicity did manage to draw the attention of both scientists and government officials in the US to the subject and so became a catalyst that activated their fusion research. An ambitious classified program was launched in 1952 and 1953. There were several experimental groups in the US that followed different arrangements of magnetic fields for confining plasma. Internal rivalry enlivened the program. Around 1954 fusion research took on the character of a crash program, with new experiments being started even before the previous ones had been made to work.

At Princeton University in New Jersey, astrophysicist Lyman Spitzer (see Fig. 5.5) invented a plasma-confinement device that he called the *stellarator*. Unlike the pinch, where the magnetic field was generated mainly by currents flowing in the plasma itself, the magnetic field in the stellarator was produced entirely by external coils. In a pinch experiment the plasma current flows around inside the torus — this is called the *toroidal* direction — and generates a magnetic field wrapped around the plasma in what is called the *poloidal* direction; see Fig. 5.6 and Box 5.2. The original idea in the stellarator had been to confine the plasma in a toroidal magnetic field. It was quickly realized that such a purely toroidal magnetic field cannot confine plasma and that it was necessary to add a twist to the field. In the first experiments this was done simply by physically twisting the whole torus into the shape of a "figure 8." Later the same effect was produced using a second set of twisted coils, the *helical winding* shown in Fig. 5.7. The stellarator has the advantage that it is capable of operating continuously. It is still considered to have the potential to be the confinement system for a fusion power plant, and active research on stellarators is being pursued in some fusion laboratories.

The fusion research program at Los Alamos in New Mexico studied toroidal pinches similar to those in the UK and also linear *theta pinches*, where a strong

Figure 5.5 ▶ Lyman Spitzer (1914–1997). As well as being a distinguished astrophysicist, Spitzer was one of the founding fathers in the field of theoretical plasma physics. The ease with which electricity is conducted through a plasma is known as the *Spitzer conductivity*. He was Professor of Astronomy at Princeton University from 1947 to 1979, and among other things he is known for being the first person to propose having a telescope on a satellite in space. His foresight is recognized by the naming of the Spitzer Space Telescope, which was launched into space by a Delta rocket from Cape Canaveral, Florida, on August 25, 2003.

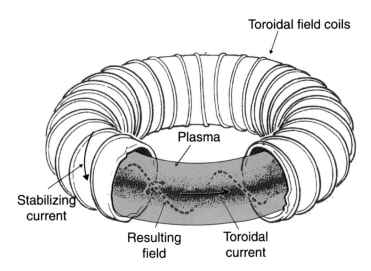

Figure 5.6 ▶ Schematic diagram of one of the first toroidal devices, the toroidal pinch. Two magnetic fields are applied: a poloidal field generated by the current flowing around in the plasma and a toroidal field produced by the external coils. The poloidal field is much stronger than the toroidal field. The combined field twists helically around the torus, as shown.

magnetic field is built up very rapidly to compress the plasma. It was thought that if this could be done sufficiently quickly, fusion temperatures might be reached before the plasma had time to escape out of the open ends. The Lawrence Livermore National Laboratory in California also built some pinches but concentrated on

> **BOX 5.2** **Toroidal Confinement**
>
> The earliest magnetic-confinement devices were developed in the UK in the late 1940s. These were *toroidal pinches* (Fig. 5.6), which attempted to confine plasma with a strong, purely poloidal magnetic field produced by a toroidal plasma current. With a sufficiently strong current, the magnetic field compresses (or *pinches* — hence the name) the plasma, pulling it away from the walls. But this arrangement proved seriously unstable — the plasma thrashed about like a snake or constricted itself like a string of sausages (Fig. 5.4). External coils adding a weak magnetic field in the toroidal direction improved stability, and further improvement was found when this toroidal field reversed direction outside the plasma — a configuration now known as the *reverse field pinch (RFP)*. Initially the field reversal occurred spontaneously, but nowadays it is induced deliberately. The potential to work at high β (see Box 5.1) would be a possible advantage of the RFP, but good energy confinement has proved elusive.
>
> The second approach to toroidal confinement is the *stellarator*, invented at Princeton in the early 1950s. This evolved as an attempt to confine fusion plasmas by means of a strong toroidal magnetic field produced by an external toroidal solenoid — without any currents in the plasma. But such a purely toroidal field cannot provide the balancing force against expansion of the plasma (this is one of the reasons the toroidal theta pinch failed). It is necessary to twist the magnetic field as it passes around the torus so that each field line is wrapped around the inside as well as the outside of the cross section (see Fig. 5.7). The coils in modern stellarators have evolved in various ways but share the same basic principle of providing a twisted toroidal magnetic field. Stellarators fell behind tokamaks in the 1960s and 1970s but now claim comparable scaling of confinement time with size. The largest experiments are the *Large Helical Device* (LHD), which came into operation at Toki in Japan in 1998, and the *W7-X* machine, under construction at Greifswald in Germany.
>
> The third, and most successful, toroidal confinement scheme is the *tokamak*, developed in Moscow in the 1960s. The tokamak can be thought of either as a toroidal pinch with very strong stabilizing toroidal field or as using the poloidal field of a current in the plasma to add the twist to a toroidal field. It is now the leading contender for a magnetically confined fusion power plant. The tokamak is described in more detail in Chapter 9.

confining plasma in a straight magnetic field by making it stronger at the ends — the so-called *mirror machine*. A similar approach using mirror machines was followed at Oak Ridge in Tennessee. Some of these linear magnetic configurations are sketched in Fig. 5.8 and outlined in Box 5.3. Although a linear system would have advantages compared to a toroidal one, in terms of being easier to build and maintain, there are obvious problems with losses from the ends. A full discussion of all the alternatives that were explored is outside the scope of this book, which will concentrate on the most successful lines that were followed.

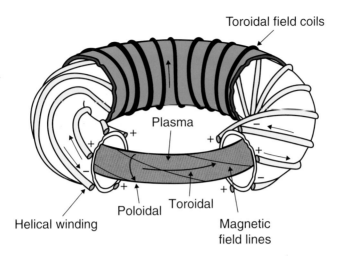

Figure 5.7 ▶ Schematic of a stellarator. The outer winding provides the toroidal field, and the inner helical winding provides the poloidal field that gives the field lines a twist, causing the magnetic field lines to spiral around inside the chamber. The toroidal field is much stronger than the poloidal field.

In the Soviet Union, the first proposal for a fusion device came in 1950 from Oleg Lavrentiev. He was a young soldier, without even a high school diploma, who was serving in the Soviet army. His proposal to confine plasma with electric rather than magnetic fields was passed on to scientists in Moscow. They concluded that electrostatic confinement would not work in this particular way, but they were stimulated to pursue magnetic confinement. This led to a strong program, initially on pinches but also expanding into the other areas, such as the open-ended mirror machines and later into the tokamak, as discussed in Chapter 9. Lavrentiev's story is remarkable. He was called to Moscow, where he finished his education with personal tuition from leading professors. His career continued at Kharkov in Ukraine, where he still works enthusiastically in fusion research.

5.3 Opening the Doors

All this research was conducted in great secrecy. Scientists in the UK and the US knew something of each other's work, but not the details. Very little was known about the work in the Soviet Union. Should fusion research be continued in secrecy or be made open? The pressure for declassification was increased in quite a dramatic way. In 1956 the Soviet leaders, Nikita Khrushchev and Nikolai Bulganin, went to the UK on a highly publicized state visit. With them was the distinguished physicist Igor Kurchatov (Fig. 5.9), who visited the Atomic Energy Research Establishment at Harwell and offered to give a lecture "On the Possibility of Producing Thermonuclear Reactions in a Gas Discharge." This was

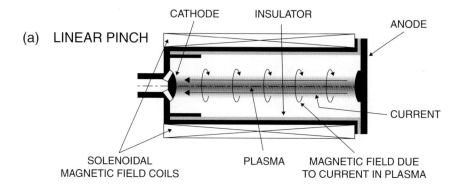

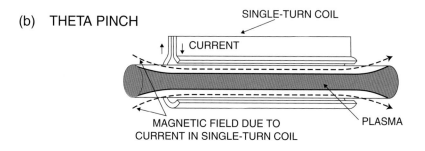

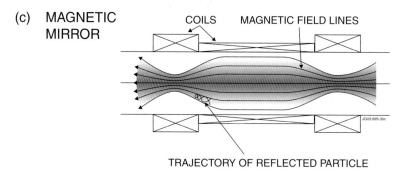

Figure 5.8 ▶ Sketch of three different linear magnetic configurations that have been considered for the confinement of hot plasma. All the linear systems had excessive losses through the ends and were abandoned.

a surprise and of great interest to the British scientists working on the "secret" fusion program. It was difficult for them to ask questions without revealing what they knew themselves, and for that they would have needed prior approval. They discovered that Soviet scientists had been following very similar lines of research into magnetic confinement as the UK and the US, concentrating on both straight

BOX 5.3 Linear Confinement

In the *linear Z-pinch*, a plasma current flowing along the axis between two end electrodes produces an *azimuthal* magnetic field that compresses (or *pinches*) the plasma away from the walls. Unfortunately the plasma rapidly becomes unstable, breaks up, and hits the wall. In fact many linear Z-pinch experiments were built to study the instabilities that had been seen in toroidal pinches. A variant of the Z-pinch, known as the *plasma focus*, was studied intensively during the 1960s. It has long since been abandoned as a serious candidate for a fusion power plant, but it continues to provide a source of dense plasma for academic studies. The linear Z-pinch survives today in the form of ultrafast, very high-current devices (sometimes using metal filaments to initiate the plasma), where it is hoped that the plasma compression and heating can take place on a faster time scale than the instabilities. The potential of these schemes as a route to magnetic confinement is doubtful, but these ultrafast pinches produce copious bursts of X-rays and are being studied as drivers for inertial-confinement fusion (Chapter 7).

The *theta pinch* is generated by a fast-rising azimuthal current in a single-turn external conductor that is wrapped around the plasma tube. This produces an axial magnetic field that compresses and heats the plasma. The SCYLLA theta pinch, developed at Los Alamos in 1958, was the first magnetic confinement system to produce really hot fusion plasmas with neutrons from thermonuclear reactions. The pulse duration of a theta pinch is very short, typically about 1 microsecond, but even on this short time scale plasma is lost by instabilities and end losses. Attempts to stopper the ends with magnetic fields or material plugs and by joining the two ends to make a toroidal theta pinch all failed. Both the Z-pinch and the theta pinch are inherently pulsed devices. Even if all the problems of end losses and instabilities were to be solved, a fusion power plant based on these configurations would be a doubtful proposition due to the large recirculating energy.

The third linear magnetic-confinement scheme has the advantage that it could run steady state. This is the *magnetic mirror machine*, where a solenoid coil produces a steady-state axial magnetic field that increases in strength at the ends. These regions of higher field, the magnetic mirrors, serve to trap the bulk of the plasma in the central lower field region of the solenoid, though ions and electrons with a large parallel component of velocity can escape through the mirrors. At low plasma density, mirror machines looked promising but had difficulties in reaching the higher densities needed for a power plant. Instabilities and collective effects caused losses that could not be overcome, in spite of adding complicated end cells to the basic mirror configuration. Development of mirror machines in the US was stopped in 1986 and programs in the former Soviet Union have been wound down due to lack of research funding. A mirror machine, GAMMA10, still operates in Tsukuba, Japan.

and toroidal pinch experiments. Kurchatov gave an elegant lecture outlining some of the main Soviet discoveries. He warned that it is possible to get production of neutrons, indicating that fusion reactions are occurring, without having a true thermonuclear reaction.

Figure 5.9 ▶ The visit of the Soviet delegation to AERE Harwell in 1956. John Cockroft, the Harwell Director, is in the left foreground; next to him is Igor Kurchatov (1903–1960). Nikita Kruschev is hidden behind Kurchatov, and Nikolai Bulganin is on Kruschev's left.

Kurchatov was one of the stars of the Soviet scientific establishment. He had supervised the building of the Soviet atom bomb, and in 1949 he laid down plans for the first Soviet atomic (fission) power plant. In 1951 he organized the first Soviet conference on controlled thermonuclear fusion, and a few months later he had set up a laboratory headed by Lev Artsimovich. Kurchatov, who died in 1960, was an advocate of nuclear disarmament; he recognized the importance of having open fusion research, and his lecture at Harwell was his first attempt at achieving it. The first open presentation of work on fusion research was at an international conference in Geneva in 1958.

5.4 ZETA

Fusion research at Harwell had expanded rapidly through the early 1950s. Bigger and ever more powerful pinch experiments were built, culminating in the ZETA machine, which started operation in 1957. ZETA was a bold venture and a remarkable feat of engineering for its day. The aluminum torus of 3 meters diameter and 1 meter bore was originally intended to contain a plasma current of 100,000 amps, but this specification was soon raised to 900,000 amps. Already in the first few weeks of operation in deuterium plasmas, currents were running at nearly 200,000 amps, and large numbers of neutrons were recorded, up to a million per pulse. This caused great excitement, but the important question was, were these neutrons *thermonuclear?* Kurchatov had already warned of the possibility that beams of ions might be accelerated to high energies and produce neutrons that could be misinterpreted as coming from the hot plasma. The distinction, though subtle, was very important and would seriously affect the way the results would extrapolate to a fusion power plant. The uncertainty could have been resolved if it had been possible to measure the plasma temperature accurately, but the techniques to do this were still in their infancy.

News of the existence of the top-secret ZETA machine and of its neutrons quickly leaked to the press. Pressure grew for an official statement from Harwell, and delays in doing so merely heightened the speculation. It was decided to publish the results from ZETA alongside papers from other British and American fusion experiments in the scientific journal *Nature* at the end of January 1958. The carefully worded ZETA paper made no claims at all about the possible thermonuclear origin of the neutrons. Harwell's director, John Cockcroft, was less cautious, and at a press conference he was drawn into saying that he was "90% certain" that the neutrons were thermonuclear. The press reported this with great enthusiasm about stories of cheap electricity from seawater. The matter was soon resolved by a series of elegant measurements on the neutrons that showed they were *not* thermonuclear. ZETA went on to make many valuable contributions to the understanding of fusion plasmas, but plans to build an even larger version were abandoned.

5.5 From Geneva to Novosibirsk

A few months after the publication of the ZETA results, the final veil of secrecy was lifted from fusion research at the Atoms for Peace conference held by the United Nations in Geneva in September 1958. Some of the actual fusion experiments were taken to Geneva and shown working at an exhibition alongside the conference. This was the first chance for fusion scientists from communist and capitalist countries to see and hear the details of each other's work, to compare notes, and to ask questions. In many cases they found that they had been working along similar lines and had made essentially the same discoveries quite independently. Above all, the conference provided an opportunity for these scientists to meet each other. Personal contacts were established that would lead to a strong tradition of international collaboration in future fusion research.

The many different magnetic configurations that had been tested could be grouped into two main categories according to whether the magnetic field was open at the ends (like the theta and Z pinches and the mirror machines) or closed into a torus (like the toroidal pinches and stellarators). Each of these categories could be further divided into systems that were rapidly pulsed and systems with the potential to be steady state. In order to meet the *Lawson criterion* for fusion, which set a minimum value for the product of density and confinement time (Box 4.3), rapidly-pulsed systems aiming to contain plasma for at most a few thousandths of a second clearly had to work at higher densities than systems that aimed to contain the plasma energy for several seconds.

Leading the field in the late 1950s in terms of high plasma temperatures and neutron production were some of the rapidly pulsed open-ended systems — in particular the theta pinches. These had impressive results because the plasma was heated by the rapid initial compression of the magnetic fields before it had time to escape or become unstable, but their potential as power plants was limited by loss of plasma through the open ends and by instabilities that developed relatively quickly. The open-ended mirror machines also suffered from end losses. The toroidal pinches and stellarators avoided end losses, but it was found that plasma escaped across the magnetic field much faster than expected. If these prevailing loss rates were scaled up to a power plant, it began to look as if it might be very difficult — perhaps even impossible — to harness fusion energy for commercial use.

In the 1960s progress in fusion research seemed slow and painful. The optimism of earlier years was replaced by the realization that using magnetic fields to contain hot plasma was very difficult. Hot plasmas could be violently unstable and, even when the worst instabilities were avoided or suppressed, the plasma cooled far too quickly. New theories predicted ever more threatening instabilities and loss processes. Novel configurations to circumvent the problems were tried, but they failed to live up to their promise. Increasingly, emphasis shifted from trying to achieve a quick breakthrough into developing a better understanding of the general properties of magnetized plasma by conducting more careful experiments with improved measurements.

Attempts to reduce the end losses from the open systems met with little success, and one by one these lines were abandoned. Toroidal systems, the stellarators and pinches, made steady progress and slowly began to show signs of improved confinement, but they were overtaken by a new contender — the *tokamak* — a configuration that had been invented in Moscow at the Kurchatov Institute. The name is an acronym of the Russian words *toroidalnaya kamera*, for "toroidal chamber," and *magnitnaya katushka*, for "magnetic coil." In 1968, just 10 years after the conference in Geneva, another major conference on fusion research was held — this time in the Siberian town of Novosibirsk. The latest results presented from the Russian tokamaks were so impressive that most countries soon decided to switch their efforts into this line. This story is continued in Chapter 9.

▶ Chapter 6

The Hydrogen Bomb

6.1 The Background

Some isotopes of uranium and plutonium have nuclei that are so close to being unstable that they fragment and release energy when bombarded with neutrons. A fission chain reaction builds up because each fragmenting nucleus produces several neutrons that can initiate further reactions. An explosion occurs if the piece of uranium or plutonium exceeds a certain *critical mass*—thought to be a few kilograms—smaller than a grapefruit. In order to bring about the explosion, this critical mass has to be assembled very quickly, either by firing together two subcritical pieces or by compressing a subcritical sphere using conventional explosives. The US developed the first atom bombs in great secrecy during World War II at Los Alamos, New Mexico. The first test weapon, exploded in New Mexico in July 1945, had a force equivalent to 21 kilotons of high explosive. A few days later, bombs of similar size devastated the Japanese cities of Hiroshima and Nagasaki.

Producing the fissile materials for such weapons was difficult and expensive and required an enormous industrial complex. Less than 1% of natural uranium is the "explosive" isotope ^{235}U and separating this from the more abundant ^{238}U is very difficult. Plutonium does not occur naturally at all and must be manufactured in a fission reactor and then extracted from the intensely radioactive waste. Moreover, the size of a pure fission bomb was limited by the requirement that the component parts be below the critical mass. Fusion did not suffer from this size limitation and might allow bigger bombs to be built. The fusion fuel, deuterium, is much more abundant than ^{235}U and easier to separate.

Even as early as 1941, before he had built the very first atomic (fission) reactor in Chicago, the physicist Enrico Fermi speculated to Edward Teller that a fission bomb might be able to ignite the fusion reaction in deuterium in order to produce an even more powerful weapon—this became known as the

Figure 6.1 ▶ Edward Teller (1908–2003). Born in Budapest, he immigrated to the US in 1935. As well as proposing the hydrogen bomb, he was a powerful advocate of the "star wars" military program.

hydrogen bomb, or *H-bomb*. These ideas were not pursued seriously until after the war ended. Many of the scientists working at Los Alamos then left to go back to their academic pursuits. Robert Oppenheimer, who had led the development of the fission bomb, resigned as director of the weapons laboratory at Los Alamos to become director of the Princeton Institute of Advanced Study and was replaced by Norris Bradbury. Edward Teller (Fig. 6.1), after a brief period in academic life, returned to Los Alamos and became the main driving force behind the development of the H-bomb, with a concept that was called the *Classical Super*.

There was, however, much soul-searching in the US as to whether it was justified to try and build a fusion bomb at all. In 1949 Enrico Fermi and Isidor Rabi, both distinguished physicists and Nobel Prize winners, wrote a report for the Atomic Energy Commission in which they said:

> Necessarily such a weapon goes far beyond any military objective and enters the range of very great natural catastrophes.... It is clear that the use of such a weapon cannot be justified on any ethical ground which gives a human being a certain individuality and dignity even if he happens to be a resident of an enemy country.... The fact that no limits exist to the destructiveness of this weapon makes its existence and the knowledge of its construction a danger to humanity as a whole. It is an evil thing considered in any light.

The debate was cut short in early 1950 by the unexpected detonation of the first Soviet fission bomb. Prompted by the suspicion that East German spy Klaus Fuchs had supplied information about US hydrogen-bomb research to the Soviet Union, President Truman ordered that the Super be developed as quickly as possible. However, no one really knew how to do this, so new fears were raised that Truman's statement might simply encourage the Soviet Union to speed up its own efforts to build a fusion bomb and, more seriously, that the Soviet scientists might already know how to do it.

6.2 The Problems

It was clear that developing a fusion weapon would be even more difficult than development of the fission bomb had been. It was necessary to heat the fusion fuel very quickly to extremely high temperatures to start the fusion reaction and to obtain the conditions for the reaction to propagate like a flame throughout all the fuel. It was recognized that the only way to do this was to set off a fission bomb and to use the tremendous burst of energy to ignite the fusion fuel. At first it was hoped that the fission bomb simply could be set off adjacent to the fusion fuel and that the heat would trigger the fusion reactions. However, calculations showed that this approach was unlikely to work. The shock wave from the fission bomb would blow the fusion fuel away long before it could be heated to a high-enough temperature to react. The problem has been likened to trying to light a cigarette with a match in a howling gale.

There were further serious problems in terms of providing the fusion fuel. A mixture of deuterium and tritium would ignite most easily, but tritium does not occur naturally and must be manufactured in a nuclear reactor. A DT fusion bomb with an explosive yield equal to that of 10 million tons of TNT (10 megatons) would require hundreds of kilograms of tritium. The rough size of the bomb can be estimated from the density of solid deuterium; it would be equivalent to a sphere about 1 meter in diameter. Smaller quantities of deuterium and tritium could be used to boost the explosive force of fission bombs or to initiate the ignition of a fusion weapon. But to manufacture tritium in the quantities needed for a significant number of pure DT bombs would require a massive production effort — dwarfing even the substantial program already under way to manufacture plutonium — and would be prohibitively expensive. Attention focused therefore on the even more difficult task of igniting explosions in pure deuterium or in a mixture of deuterium and lithium. The lithium could be converted into tritium *in situ* using neutrons that would be produced as the fuel burned.

Teller's original idea had been to explode a small fission bomb at one end of a long cylinder of frozen deuterium. The basic idea was that the fission bomb would heat the fusion fuel sufficiently to ignite a fusion reaction in the deuterium. A refinement was to use a small amount of tritium mixed with the deuterium to help start the ignition. If this worked, in principle there would be no limit to the strength of the explosion, which could be increased just by making the deuterium cylinder longer. However, calculations by Stanislaw Ulam, another Livermore scientist, had cast doubts on its feasibility and shown that unrealistically large quantities of tritium would be required. Teller proposed a number of new designs in late 1950, but none seemed to show much promise.

Early in 1951, Ulam made an important conceptual breakthrough, and Teller quickly refined the idea. This followed an idea known as *radiation implosion* that had first been broached in 1946 by Klaus Fuchs before he was arrested for giving atomic secrets to the Soviet Union. Most of the energy leaves the fission trigger as X-rays. Traveling at the speed of light, the X rays can reach the nearby fusion fuel almost instantaneously and be used to compress and ignite it before it is blown apart by the blast wave from the fission explosion, which travels only at the speed of sound. This is analogous to the delay between seeing a flash of lightning and hearing the sound of the thunder, though the much smaller distances in the bomb reduce the delay to less than a millionth of a second. A second important requirement is that the radiation from the fission bomb needs to compress the fusion fuel before it is heated to the high temperature at which it will ignite. It is much easier to compress a gas when it is cold than when it is hot.

The technique for compressing the fusion fuel has since become known as the *Teller–Ulam configuration*, and it is shown schematically in Fig. 6.2. Although many of the details remain secret, in 1989 there was a partial declassification of the history of American hydrogen bomb development, and the basic principles are known. A description of the bomb is now included in the *Encyclopedia Britannica*. Like all technical achievements, once it is known that it can be done, it is much easier to work out how to do it.

The fission bomb trigger is set off at one end of a cylindrical casing in which the fusion fuel is also contained. The fusion fuel is thought to be in the form of a cylinder surrounding a rod of plutonium, and a layer of very dense material — usually natural uranium or tungsten — surrounds the fuel itself. The X-ray radiation from the fission bomb is channeled down the radial gap between the outer casing and the fusion fuel cylinder. The gap is filled with plastic foam that is immediately vaporized and turned into hot plasma. The plasma is transparent to the X-rays, allowing the inner surface of the cylindrical casing and the outer surface of the dense layer surrounding the fusion fuel to be heated quickly to very high temperatures. As the outer surface of the layer surrounding the fuel vaporizes, it exerts a high inward pressure — rather like an inverted rocket engine.

Enormous pressures are generated instantaneously — several billion times atmospheric pressure — and the fuel is compressed to typically 300 times its normal density. It is salutary to realize that the explosive force released by the fission trigger, enough to destroy an entire city, is being used briefly to squeeze a few kilograms of fuel! The compression and the neutrons from the fission bomb cause the plutonium rod down the middle of the fusion fuel to become critical and to explode — in effect a second fission bomb goes off. This explosion rapidly heats the already compressed fusion fuel to the temperature required

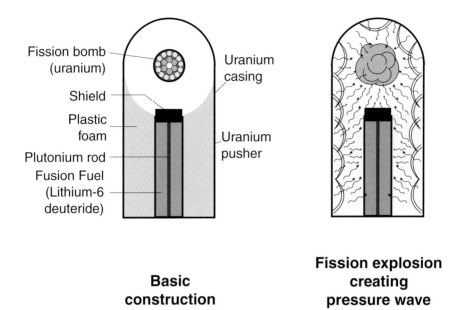

Figure 6.2 ▶ Schematic diagram of the elements of an H-bomb. A fission explosion is first triggered by high explosive. This explosion is contained inside a heavy metal case. The radiation from the fission bomb causes the implosion and heating of the fusion fuel and sets off the fusion bomb.

for the fusion reactions to start. Once ignited, the fusion fuel burns outward and extremely high temperatures — up to 300 million degrees — are reached as almost the whole of the fuel is consumed. Some recent reports suggest that the design has been refined so much that the plutonium "spark plug" is no longer needed.

The first thermonuclear bomb test, code named Ivy–Mike, took place at Eniwetok Atoll in the Pacific Ocean in November 1952 and achieved a yield equivalent to 10 megatons (Fig. 6.3). It is estimated that only about a quarter of this came from fusion and the rest from fission induced in the heavy uranium case. This was a test of the principle of the compression scheme rather than a deployable weapon. The fusion fuel was liquid deuterium, which had to be refrigerated and contained in a large vacuum flask. A massive casing of uranium and steel surrounded this. The whole device weighed over 80 tons — hardly a weapon that could be flown in an airplane or fired from a missile.

In order to make a more compact weapon, a version using solid fusion fuel, *lithium deuteride*, was developed. In this case tritium is produced in the course of the reaction by neutron bombardment of the lithium, and the principal fusion reaction is between deuterium and tritium. Lithium deuteride, a chemical compound of lithium and deuterium, is a solid at room temperature, and this approach

Figure 6.3 ▶ Photograph of the explosion of the first fusion bomb over Eniwetok Atoll in November 1952.

obviates the necessity of the complex refrigeration system. The US tested this concept in the spring of 1954, with a yield of 15 megatons.

6.3 Beyond the "Sloyka"

The Soviet Union had also started to think about fusion weapons in the late 1940s and recognized the same difficulties as the Americans in igniting explosions in pure deuterium. The first Soviet H-bomb test, in 1953, consisted of alternating layers of lithium deuteride and uranium wrapped around the core of a fission bomb. This concept, by Andrei Sakharov (Fig. 6.4), was called the *sloyka*—a type of Russian layer cake. The principle was to squeeze the fusion fuel between the exploding core and the heavy outer layers. Teller had apparently considered something rather similar that he called the "alarm clock." This was not a true H-bomb, in that most of the energy came from fission, and its size was limited to less than a megaton.

Figure 6.4 ▶ Andrei Sakharov (1921–1989). He made many contributions to the nuclear fusion program, and, along with Igor Tamm, he is credited with the invention of the tokamak. He played a crucial role in the Soviet atomic weapons program until 1968, when he published his famous pamphlet, *Progress, Peaceful Coexistence and Intellectual Freedom*. He was awarded the Nobel Peace Prize in 1975 but was harassed by the Soviet authorities for his dissident views.

Andrei Sakharov is also credited with independently conceiving the idea of a staged radiation implosion, similar to that of Teller and Ulam, which led to the first true Soviet H-bomb test in November 1955. It is thought that most subsequent thermonuclear weapons, including those developed by other countries, have been based on principles similar to the Teller–Ulam configuration that has been described here, although of course most of the details remain secret.

▶ Chapter 7

Inertial-Confinement Fusion

7.1 Mini-Explosions

The inertial-confinement route to controlled-fusion energy is based on the same general principle as that used in the hydrogen bomb — fuel is compressed and heated so quickly that it reaches the conditions for fusion and burns before it has time to escape. The inertia of the fuel keeps it from escaping — hence the name *inertial-confinement fusion* (ICF).

Of course, the quantity of fuel has to be much smaller than that used in a bomb so that the energy released in each "explosion" will not destroy the surrounding environment. The quantity of fuel is constrained also by the amount of energy needed to heat it sufficiently quickly. These considerations lead to typical values of the energy that would be produced by each mini-explosion that lie in the range of several hundred million joules. To put this into a more familiar context, 1 kilogram of gasoline has an energy content of about 40 million joules, so each mini-explosion would be equivalent to burning about 10 kilograms of gasoline. Due to the much higher energy content of fusion fuels, this amount of fusion energy can be produced by burning a few milligrams of a mixture of deuterium and tritium — in solid form this is a small spherical pellet, or *capsule*, with a radius of a few millimeters. An inertial fusion power plant would have a chamber where these mini-explosions would take place repeatedly in order to produce a steady output of energy. In some ways this would be rather like an automobile engine powered by mini-explosions of gasoline. The main sequence of events is shown schematically in Fig. 7.1.

The general conditions for releasing energy from fusion that were discussed in Chapter 4 are essentially the same for inertial confinement as for magnetic confinement. To recap, for fusion reactions to occur at a sufficient rate requires a temperature in the region of 200 million degrees, and to obtain net energy production the fuel density multiplied by the confinement time has to be larger than

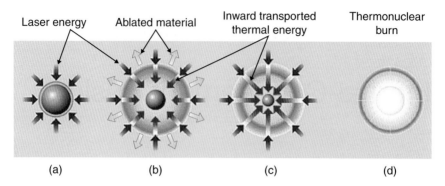

Figure 7.1 ▶ The four stages of a fusion reaction in an inertial-confinement capsule: (a) Laser heating of the outer layer. (b) Ablation of the outer layer compresses the capsule. (c) The core reaches the density and temperature for ignition. (d) The fusion reaction spreads rapidly through the compressed fuel.

about 10^{21} nuclei per cubic meter multiplied by seconds. In magnetic confinement, the time involved is of the order of seconds and the density of the plasma is in the range 10^{20}–10^{21} nuclei per cubic meter — many times less dense than air. In inertial-confinement fusion, the time involved is a few tenths of a billionth of a second and the density of the plasma has to reach 10^{31} nuclei per cubic meter — many times more dense than lead. The term *ignition* is used in both magnetic- and inertial-confinement fusion, but with different meanings. Magnetic confinement aims at reaching steady-state conditions, and ignition occurs when the alpha particle heating is sufficient to maintain the whole plasma at a steady temperature. Inertial-confinement fusion is inherently pulsed, and ignition occurs when the fuel capsule starts to burn outward from a central hot spot.

It is convenient to express the ignition criterion for inertial confinement in terms of the radius of the capsule multiplied by its density. Knowing the temperature of the plasma allows us to calculate the expansion velocity of the capsule and thus to convert the time in the conventional criterion into a radius (Box 7.1). The minimum value of the product of density and radius depends on the efficiency of converting fusion energy output into effective capsule heating. In order to reach the required value, it is necessary to compress the solid fuel to very high density (Box 7.2).

Usually solids and liquids are considered to be incompressible. The idea that they can be compressed is outside our everyday experience, but it can be done if the pressure on the capsule is large enough. Experimental results show that it is possible to compress deuterium capsules to densities more than 1000 times their normal density.

The key to compressing a capsule is to heat its surface intensely so that it evaporates (or *ablates*) very rapidly (Fig. 7.2). Hot gases leaving the surface act like those leaving a rocket engine, and they apply large forces to the capsule.

BOX 7.1 — Conditions for Inertial Confinement

The conditions required for ICF are determined by the requirement that the fusion energy produced in one pulse exceeds the energy invested in heating the fuel to the ignition temperature. The fusion energy (17.6 MeV per reaction) is extracted after each pulse and converted into electricity, and some of this energy has to be used to heat the fuel for the next pulse. For breakeven, with T in keV as in Box 4.3,

$$\frac{1}{4} n^2 \overline{\sigma v} 17.6 \times 10^3 k_e \tau > 3nkT$$

$$n\tau > 6.82 \times 10^{-4} (T/\overline{\sigma v}) \epsilon^{-1} \text{ m}^{-3} \text{ s}$$

Note that τ is the pulse duration (the capsule burn time) and ϵ is the overall efficiency of converting fusion energy in the form of heat into effective capsule heating. This is the expression that John Lawson derived in the 1960s — and he was thinking in terms of pulsed magnetic confinement fusion, so he assumed that all the fusion energy would be converted from heat into electricity with $\epsilon \approx 0.33$ and used for ohmic heating of the next pulse. At $T \approx 30$ keV, this gives the well-known expression $n\tau > 1 \times 10^{20}$ m^{-3} s. However, $\epsilon \ll 0.33$ for ICF because of the need to convert heat into electricity and electricity into driver energy and then to couple the driver energy into effective fuel heating.

An ICF fuel capsule is compressed initially to radius r (Box 7.2) and heated until it reaches the density and temperature for fusion. Then it starts to burn and expands in radius. The plasma density ($n \propto 1/r^3$) and fusion power ($P_F \propto n^2$) fall rapidly as the capsule expands, and we assume that the capsule stops burning when its radius has expanded by 25%, so $\tau \approx r/4v_i$. The expansion velocity of the capsule is determined by the ions (the electrons are held back by the heavier ions) and $v_i = 2 \times 10^5 T^{0.5}$ m/s (for a 50:50 mixture of deuterium and tritium). Thus,

$$nr > 545 \left(T^{1.5}/\overline{\sigma v} \right) \epsilon^{-1} \text{ m}^{-2}$$

The optimum is at $T \approx 20$ keV, where $T^{1.5}/\overline{\sigma v} \approx 2.1 \times 10^{23}$ keV$^{1.5}$ m^{-3} s and

$$nr > 1.15 \times 10^{26} \epsilon^{-1} \text{ m}^{-2}$$

This expression can be written in terms of the mass density $\rho = n \times 4.18 \times 10^{-27}$ kg m^{-3}. Thus,

$$\rho r > 0.48 \epsilon^{-1} \text{ kg m}^{-2} \quad (\text{or } 0.048 \epsilon^{-1} \text{ g cm}^{-2})$$

This is the condition for breakeven (when the fusion energy output is just sufficient to heat the next capsule). A fusion power plant has to produce a net output of power, and the condition is more stringent, as discussed in Box 11.3.

> **BOX 7.2** **Capsule Compression**
>
> If the capsule has radius r_0 before compression and is compressed by a factor c to $r = r_0/c$ before igniting, the compressed particle density will be $n = c^3 n_0$ and the mass density $\rho = c^3 \rho_0$. Using the results from Box 7.1, taking $\rho_0 \approx 300$ kg m^{-3} as the uncompressed density of a mixture of solid deuterium and tritium and expressing r_0 in millimeters and ϵ as a percentage, the condition for the minimum compression is
>
> $$c^2 > 160(r_0 \epsilon)^{-1}$$
>
> For example a capsule with uncompressed radius of 1 mm and conversion efficiency of 1% would require compression $c \approx 13$ to reach breakeven. Clearly it is desirable to work at high conversion efficiency, but this is limited by the driver technology.
>
> The maximum capsule size is constrained by the maximum explosion that can be safely contained within the power plant. The fusion energy released by burning a capsule with uncompressed radius r_0 (in millimeters) is $4.25 \times 10^8 r_0^3$ J $= 0.425 r_0^3$ GJ. A 1-mm-radius DT capsule is equivalent to about 10 kg of high explosive.
>
> The calculations in this box and Box 7.1 contain significant simplifications. Ideally only the core of the compressed capsule needs to be heated to ignition temperature, and, once the core ignites, the fusion reaction propagates rapidly through the rest of the compressed fuel. This reduces the energy invested in heating the capsule, but account also has to be taken of the energy needed for compression and of incomplete burn-up. Sophisticated numerical computer models are used to calculate the behavior of compressed capsules.

The thrust developed for a brief instant in capsule compression is 100 times that of a space shuttle launcher! If the capsule surface is heated on just one side, the force of the evaporating gas will accelerate the capsule in the opposite direction, just like a rocket. However, if the capsule is heated uniformly from all sides, these forces can generate pressures about 100 million times greater than atmospheric pressure and compress the capsule.

7.2 Using Lasers

Until the beginning of the 1960s there appeared to be no power source large enough to induce inertial-confinement fusion in a controlled way. The development of the laser suggested a method of compressing and heating fuel capsules on a sufficiently fast time scale. A laser is a very intense light source that can be focused down to a small spot size and turned on for just the right length of time to compress and heat the capsule. The principle of the laser (Box 7.3) was proposed

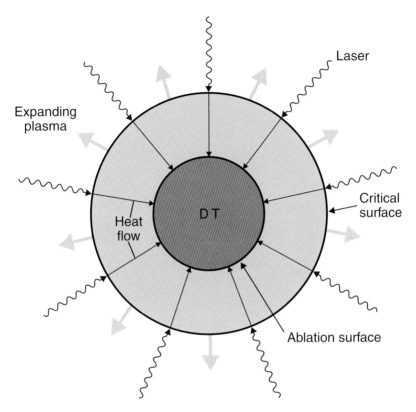

Figure 7.2 ▶ The ablation of a capsule by incident laser light creates a plasma expanding outward that applies a force in the opposite direction, compressing the capsule. Laser light cannot penetrate the dense plasma beyond a layer known as the *critical surface*. The laser energy is absorbed close to the critical surface and carried in to the *ablation surface* by heat conduction.

by Arthur Schalow and Charles Townes at the Bell Telephone Laboratories in 1958, and the first working laser was built by Theodore Maimon in 1960. This was just at a time when magnetic-confinement fusion had reached something of an impasse, so laser enthusiasts saw an alternative approach to fusion. As early as 1963, Nicolai Basov (Figure 7.5) and Alexandr Prokhorov at the Lebedev Institute in Moscow put forward the idea of achieving nuclear fusion by laser irradiation of a small target. Some laser physicists thought that they could win the race with magnetic confinement to obtain energy *breakeven*, where energy produced equals the initial heating energy.

The energy of the early lasers was too small, but laser development was proceeding rapidly, and it seemed likely that suitable lasers would soon be available. Basov and colleagues in Moscow reported the generation of thermonuclear neutrons from a plane target under laser irradiation in 1968, and the Limeil group in France demonstrated a definitive neutron yield in 1970. The expertise developed in the design of the hydrogen bomb included detailed theoretical models that could be run on computers to calculate the compression and heating of the fuel. This close overlap with nuclear weapons had the consequence that much of

---

BOX 7.3   The Laser Principle

A laser (Fig. 7.3) produces coherent light at a well-defined wavelength with all the photons in phase. To produce laser light it is necessary to have a suitable medium in which atoms or molecules can be excited to a *metastable* level above the ground state. The atoms are excited typically by a *flash tube*, which when energized emits an intense burst of incoherent light. Some photons from the flash tube have the right wavelength to excite the atoms in the laser. Having accumulated a sufficient population of excited atoms in the metastable state, they can all be induced to decay at the same time by a process called *stimulated emission*. A single photon, of the same wavelength that is emitted, causes one excited atom to decay, producing a second photon and an avalanche process develops. The emitted photons have the same phase and propagation direction as the photon triggering the emission. Mirrors are placed at either end of the lasing medium so that each photon can be reflected back and forth, causing all the excited atoms to decay in a time that is typically nanoseconds. The mirrors form an optical cavity, resonant with the desired wavelength of the laser, that can be tuned by adjusting the separation between the mirrors.

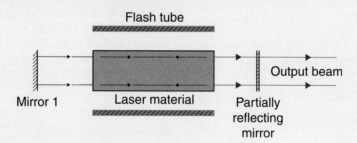

Figure 7.3 ▶   Schematic of a simple laser system.

Suitable lasing media, where metastable excited states can be produced, include solids, liquids, gases, and plasmas. Normally the excitation of the metastable state is a two-stage process, where the atom is initially excited to a higher-energy state that then decays to the metastable state (Fig. 7.4). The laser light is coherent and can be focused to a very small spot, producing very high-power densities. The first lasers tended to be in the red or infrared wavelengths, but suitable media have been found that enable lasers in the visible, ultraviolet, and even X-ray wavelengths to be produced. It is also possible to double or triple the frequency of the laser light using special optical materials that have nonlinear properties. This is particularly important for ICF to bring the wavelength of the light from a neodymium laser (1.053 μm) into a part of the spectrum (0.351 μm) where the compression is more effective.

BOX 7.3 (continued)

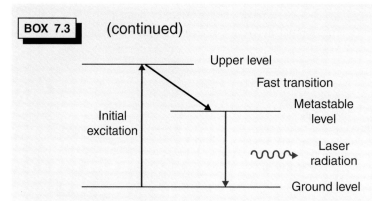

Figure 7.4 ▶ The energy levels of an atom in which lasing is produced by a two-stage process. Excitation to the upper level, followed by prompt decay to a metastable level, sets up the conditions for stimulated emission.

the inertial-confinement research remained a closely guarded secret, very reminiscent of the early days of magnetic confinement. Some details were released in 1972 when John Nuckolls (Fig. 7.6) and his collaborators at the Lawrence Livermore National Laboratory in California outlined the principles in their landmark paper published in the international science journal *Nature*. Optimists predicted that inertial-confinement fusion might be achieved within a few years, but it soon became clear that it was not so easy.

Estimates of the amount of laser energy required to compress and heat a fusion capsule have increased severalfold since the early days, and present calculations show that at least 1 million joules will be required. This requires very advanced types of laser, and the most suitable ones presently available use a special sort of glass containing a rare element called *neodymium*. The laser and its associated equipment fill a building larger than an aircraft hanger. These lasers consist of many parallel beams that are fired simultaneously and focused uniformly onto the small fusion capsule. Typical of these types of laser was the NOVA facility at the Lawrence Livermore National Laboratory in the US (Fig. 7.7).

NOVA came into operation in 1984 and had 10 beams. It was capable of producing up to 100,000 (10^5) joules in a burst of light lasting a billionth (10^{-9}) of a second (Fig. 7.8). For that brief instant its output was equivalent to 200 times the combined output of all the electricity-generating plants in the US. At the present time the most powerful laser operating in the US is the OMEGA facility at the University of Rochester. Large lasers have also been developed at the Lebedev Institute in Russia, at Osaka University in Japan, at Limeil in France, and at other laboratories around the world. New lasers with state-of-the-art technology (Fig. 7.9) and energies more than 10 times that of existing

Figure 7.5 ▶ Nicolai Basov (1922–2001), Nobel Laureate in Physics jointly with Charles Townes and Alexandr Prokhorov in 1964 for their development of the field of quantum electronics, which led to the discovery of the laser. Basov was the leader of the Soviet program on inertial-confinement fusion for many years.

Figure 7.6 ▶ John Nuckolls (b. 1931) was the lead author of the landmark paper on inertial-confinement fusion in 1972 and a strong proponent of ICF in the US. He was director of the Lawrence Livermore National Laboratory from 1988 to 1994.

Section 7.2 **Using Lasers** 77

Figure 7.7 ▶ The NOVA laser facility at the Lawrence Livermore National Laboratory. The central sphere is the target chamber, and the large pipes contain the optical systems bringing in the multiple laser beams.

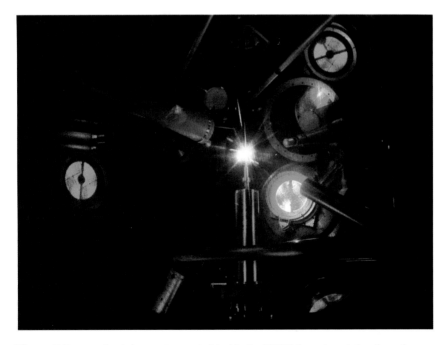

Figure 7.8 ▶ A miniature star created inside the NOVA laser target chamber when an experimental capsule is compressed and heated by laser beams.

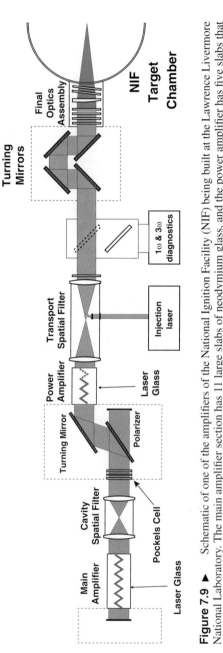

Figure 7.9 ▶ Schematic of one of the amplifiers of the National Ignition Facility (NIF) being built at the Lawrence Livermore National Laboratory. The main amplifier section has 11 large slabs of neodymium glass, and the power amplifier has five slabs that are pumped by arrays of flash tubes (not shown). Laser light enters from a master oscillator and passes four times through the main amplifier and twice through the power amplifier sections before being directed to the target chamber. The infrared light from the laser is converted into ultraviolet light in the final optics assembly before it enters the target chamber. NIF will have 192 of these amplifiers in parallel to meet its requirement of 1.8 million joules.

facilities are under construction at Livermore in the US and at Bordeaux in France (Section 7.4).

A neodymium laser produces an intense burst of light in the infrared part of the spectrum — just outside the range of visible light that can be seen by the human eye. However, this wavelength is not ideal for capsule compression. When the outer surface of a capsule is evaporated, it forms a high-density plasma that screens the interior of the capsule. Laser light in the infrared is reflected by this plasma screen, while light at a shorter wavelength, in the ultraviolet part of the spectrum, can penetrate deeper. In order to take advantage of the deeper penetration, the wavelength of the neodymium laser light is shifted from the infrared to the ultraviolet using specialized optical components, but some energy is lost in the process. Other types of laser, for example, using krypton fluoride gas, are being developed to work directly in the optimum ultraviolet spectral region, but at present they are less powerful than neodymium lasers.

One of the problems inherent in compressing capsules in this way is that they tend to become distorted and fly apart before they reach the conditions for fusion. These distortions, known as *instabilities*, are a common feature of all fluids and plasmas. One particular form of instability (known as the *Rayleigh–Taylor instability*) that affects compressed capsules was first observed many years ago in ordinary fluids by the British physicist Lord Rayleigh. It occurs at the boundary between two fluids of different density and can be observed if a layer of water is carefully floated on top of a less dense fluid such as oil. As soon as there is any small disturbance of the boundary between the water and the oil, the boundary becomes unstable and the two fluids exchange position so that the less dense oil floats on top of the denser water. A similar effect causes the compressed capsule to distort unless its surface is heated very uniformly. A uniformity of better than 1% is called for, requiring many separate laser beams.

An ingenious way of obtaining a high degree of power uniformity was suggested at Livermore around 1975, although the details remained secret until many years later. The capsule is supported inside a small metal cylinder that is typically a few centimeters across and made of a heavy metal such as gold, as in Fig. 7.10. This cylinder is usually called a *hohlraum*, the German word for "cavity." The laser beams are focused through holes onto the interior surfaces of this cavity rather than directly onto the capsule. The intense laser energy evaporates the inner surface of the cavity, producing a dense metal plasma. The laser energy is converted into X-rays, which bounce about inside the hohlraum, being absorbed and reemitted many times, rather like light in a room where the walls are completely covered by mirrors. The bouncing X-rays strike the capsule many times and from all directions, smoothing out any irregularities in the original laser beams. X-rays can penetrate deeper into the plasma surrounding the heated capsule and couple their energy more effectively than longer-wavelength light. Some energy is lost in the conversion, but the more uniform heating compensates for this. This approach is known as *indirect drive*, in contrast to the *direct-drive* arrangement, where the laser beams are focused directly onto the capsule. Both approaches (Fig. 7.11) are being studied in inertial-confinement experiments.

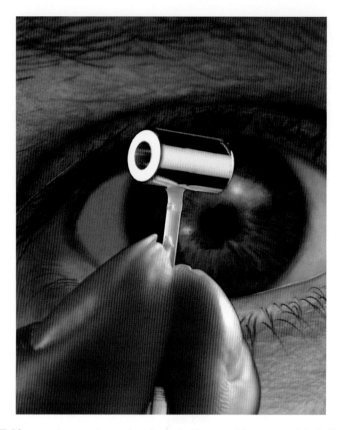

Figure 7.10 ▶ An experimental cavity, or *hohlraum*, of the type used for indirect-drive experiments. Intense laser light is shone into the cavity through the holes and interacts with the wall, creating intense X-rays. These X-rays heat the capsule, causing ablation, compression, and heating.

An equally important part of the inertial-confinement process is the design of the capsules themselves (see Box 7.4). Capsules typically consist of a small plastic or glass sphere filled with tritium and deuterium — more sophisticated targets use multiple layers of different materials with the objective of making the processes of ablation and compression more efficient.

It is important to avoid heating the capsule core too strongly during the early part of the compression phase because much more energy is then needed to compress the hot plasma. Ideally the core of the capsule should just reach the ignition temperature at the time of maximum compression. Once the core of the capsule starts to burn, the energy produced by the fusion reactions will heat up the rest of the capsule and the fusion reaction will spread outward. A more sophisticated scheme has been proposed by Max Tabak and colleagues at Livermore, where an ultrafast laser is fired to heat and ignite the core after it has been compressed.

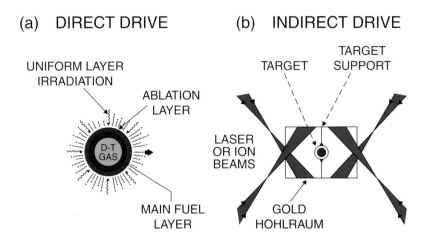

Figure 7.11 ▶ Comparison of the principles of direct and indirect drive. Uniform irradiation is produced by many laser beams in the direct-drive case and by the production of X-rays at the wall of the hohlraum in the indirect case. The targets are similar in size in the two cases, but the direct drive has been shown enlarged to illustrate the typical structure.

This technique, known as *fast ignition* (Box 7.5), may reduce the overall energy requirements for the driver.

7.3 Alternative Drivers

The big neodymium lasers presently at the forefront of ICF research are very inefficient — typically less than 1% of electrical energy is converted into ultraviolet light, and there is a further loss in generating X-rays in the hohlraum. A commercial ICF power plant will require a much higher driver efficiency (see Chapter 11), otherwise all its output will go into the driver. Using light-emitting diodes instead of flash tubes to pump a solid-state laser would improve the efficiency and permit the rapid firing rate needed for a power plant — but diode systems are extremely expensive at the moment, and costs would have to fall dramatically. Other types of lasers might also be developed, and ICF research is looking at alternative drivers using beams of energetic ions and intense bursts of X-rays.

Low-mass ions, such as lithium, would be attractive as drivers because they can be produced efficiently with energy in the range needed to compress and heat a capsule. Systems to accelerate and focus intense beams of light ions have been developed at Sandia National Laboratories, but it has proved difficult to achieve adequate focusing, and the power intensities required for an ICF driver have not

BOX 7.4 Capsule Design

The ICF target capsule is generally a spherical shell filled with low-density DT gas (<1.0 mg cm^{-3}), shown schematically in Fig. 7.1a. The outer layer is usually a plastic shell, which forms the ablator, and the inner layer of frozen deuterium–tritium (DT) forms the main fuel. The driver energy is deposited rapidly on the ablator layer, which heats up and evaporates. As the ablator evaporates outward, the rest of the shell is forced inward to conserve momentum. The capsule behaves as a spherical, ablation-driven rocket. The compression is achieved by applying a laser pulse that is carefully shaped in time so that it starts off at a low intensity and then increases to a maximum. A larger, thinner shell that encloses more volume can be accelerated to higher velocity than a thicker shell of the same mass, but the maximum ratio of capsule radius to shell thickness that can be tolerated is limited by Rayleigh–Taylor instabilities. The peak implosion velocity determines the minimum energy and mass required for ignition of the fusion fuel. The smoothness and uniformity of the capsule surface are important factors in determining the maximum compression that can be reached. The typical surface roughness must be less than 100 nanometers (10^{-7} m).

The reference design target for the National Ignition Facility (NIF) is a 2.22-mm-diameter capsule with an outer plastic ablator shell. The DT fuel solidifies on the inside wall of the plastic shell, leaving the center filled with DT gas at 3×10^{-4} g cm^{-3}, corresponding to the vapor pressure of the cryogenic layer. The target is predicted to implode at a velocity of 4×10^7 cm s^{-1} to a peak fuel density of 1.2 kg cm^{-3}, corresponding to $\rho r = 1.5$ g cm^{-2}.

In its final, compressed state, the fuel reaches pressures up to 200 gigabars and ideally consists of two regions, a central hot spot containing 2–5% of the fuel and a cooler main fuel region containing the remaining mass. Ignition occurs in the central region, and a thermonuclear burn front propagates outward into the main fuel. Double-shell capsules with an inner gold shell are also being investigated but so far have not performed as well as expected.

---

been obtained. Another difficulty is that the focusing elements have to be a very short distance from the target, while in order to survive the effects of the fusion mini-explosion these elements should be at least several meters away.

Heavy-ion beams, such as xenon, cesium, and bismuth, are also being studied as drivers. Heavy ions need to be accelerated to a much higher energy than light ions, about 10,000 MeV, but the current would be smaller, and focusing to a small spot would be easier. However, the beam current requirements are still many orders of magnitude beyond what has been achieved in existing high-energy accelerators. The heavy-ion approach to ICF is certainly interesting and promising, but testing the basic physics and technology will require the construction of a very large accelerator.

BOX 7.5 Fast Ignition

The conventional scheme for inertial confinement uses the same laser for both compression and heating of the capsule. Fast ignition uses one laser to compress the plasma, followed by a very intense fast-pulsed laser to heat the core of the capsule after it is compressed. Separating the compression and heating into separate stages could lead to a reduction in the overall driver energy requirements.

Laser light cannot penetrate through the dense plasma that surrounds the compressed capsule core, and the key to fast ignition is that a very intense, short-pulsed laser might be able to bore its way in. Laser light focused onto a very small spot with intensity as high as 10^{18}–10^{21} W cm^{-2} produces a relativistic plasma with beams of very high-energy electrons moving close to the speed of light, so they can in fact keep up with the laser beam. The basic idea is that the relativistic electron beam punches a channel through the compressed fuel and heats the core to ignition. Relativistic electron beams are produced when laser beams are focused onto target foils, but instabilities might break up these beams and prevent them from reaching the core of a compressed capsule. This is currently an active area of experimental and theoretical research.

An ingenious way around the problem has been suggested. A small cone is inserted into the fusion capsule to keep a corridor free of plasma during the compression along which the fast ignition laser pulse can propagate to the center of the compressed fuel. This idea was tested successfully in 2001 by a Japanese/UK team using the Gekko XII laser at Osaka University, Japan, to compress a capsule and a 10^{14} W (100 *terawatts*) fast laser to heat it. Fast lasers at least 10 times more powerful — in the *petawatt* (10^{15} W) range will be needed to approach breakeven and are being developed in several laboratories. If these developments are successful, then it is probable that fast ignition will be an option for the NIF and Laser Megajoule (LMJ) facilities.

The Sandia *Z facility* which had been built to study light-ion beams, is now testing a different configuration that might drive ICF by producing an intense burst of X-rays from an ultrafast pinch discharge in an array of thin tungsten wires. Electrical energy is stored in a large bank of capacitors and released through a special circuit that delivers a current of 20 million amps in a submicrosecond pulse (Fig. 7.12). The wires vaporize, forming a metal plasma that is accelerated by the enormous electromagnetic forces to implosion velocities as high as 750 kilometers per second (7.5×10^5 ms^{-1}). In one arrangement, known as a *dynamic hohlraum*, the tungsten wire array surrounds a high-gain ICF capsule (Fig. 7.13) in a block of low-density plastic foam. The imploding wire array forms a metal plasma shell that acts as the hohlraum wall, trapping the X-ray radiation that is formed when the plasma interacts with the foam and ablates the outer shell of the capsule.

Figure 7.12 ▶ An open-shutter photo of the *Z facility* at Sandia National Laboratories in Albuquerque showing the "arcs and sparks" produced when the energy storage bank is fired.

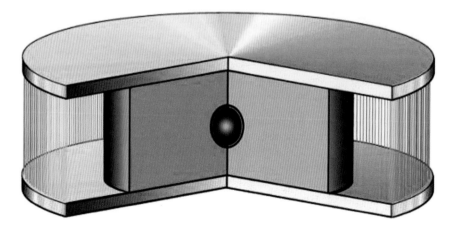

Figure 7.13 ▶ Schematic of the dynamic hohlraum concept. The ICF capsule is surrounded by low-density plastic foam at the center of the array of thin tungsten wires. An ultrafast pinch discharge drives a metal plasma inward at very high velocity, vaporizing the foam and forming its own hohlraum around the capsule.

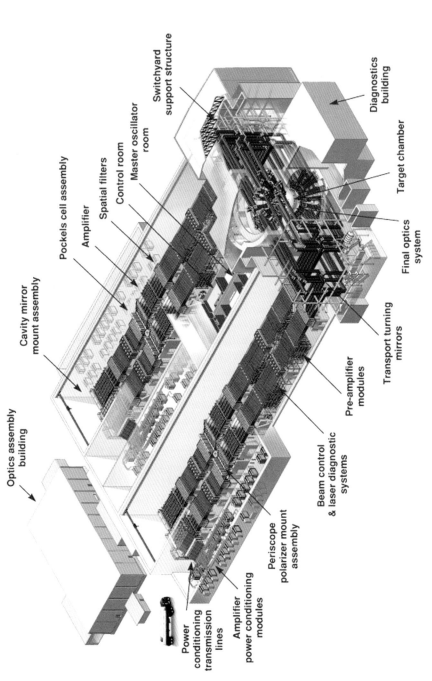

Figure 7.14 ▶ Schematic of the National Ignition Facility (NIF) being built at the Lawrence Livermore National Laboratory and due to be completed in 2010. The building contains the lasers and all the support equipment to deliver a laser pulse with an energy of 1.8 million joules (MJ) and a peak power of 500 million million watts (TW) onto the ICF capsule in the target chamber.

7.4 The Future Program

A demonstration of ignition and burn propagation is the next important goal of the inertial-confinement program. Two large laser facilities are under construction: the *National Ignition Facility (NIF)* at the Lawrence Livermore National Laboratory in California and the *Laser Megajoule (LMJ)* facility near Bordeaux, in France. Both facilities will be used for ICF and for a variety of basic science studies, although their construction was motivated to maintain scientific expertise in nuclear weapon technology following the introduction of the treaty banning nuclear weapons testing.

The NIF and the LMJ facility are based on neodymium glass lasers, whose output in the infrared at 1.056 micrometers will be converted to the ultraviolet at 0.35 micrometers. The energy is specified to be about 1.8 million joules, a big step when compared with the earlier NOVA laser. The building for the new NIF laser system (Fig. 7.14) is as large as a football stadium. The LMJ facility will be very similar, and many of the laser components are being developed jointly by the two projects.

NIF proposes to use both the direct-drive and the indirect-drive approaches discussed earlier in this chapter. To meet the stringent requirements of direct drive, it is being built with 192 laser beams, and optical smoothing techniques are being developed that will allow very uniform heating. The program for LMJ will concentrate on indirect drive and will have fewer but more energetic beams (60 beams, each of 30 kilojoules) to arrive at the total energy of 1.8 megajoules. A test beam is already under construction, and experiments with an energy of 60 kilojoules are planned. The full system of 60 beams will be completed in 2010. The target chamber will be 10 meters in diameter and is designed to absorb up to 16 megajoules of neutrons, 3 megajoules of X-rays, and 3 megajoules of particle debris.

When these new facilities come into operation, they will allow the ICF concept to be tested with both direct and indirect drive in conditions close to energy breakeven. The parallel development of ideas such as fast ignition and alternative drivers holds the key to the prospects of developing ICF as a potential source of energy.

▶ Chapter 8

False Trails

8.1 Fusion in a Test Tube?

The world was taken by surprise in March 1989 when Stanley Pons and Martin Fleischmann held a press conference at the University of Utah in Salt Lake City to announce that they had achieved fusion at room temperature "in a test tube." Although Pons and Fleischmann were not experts in fusion, they were well known in their own field of electrochemistry, and Martin Fleischmann was a Fellow of the prestigious Royal Society. Their claims quickly aroused the attention of world media, and "cold fusion" made all the headlines. It seemed that these two university scientists with very modest funding and very simple equipment had achieved the goal sought for decades by large groups with complex experiments and budgets of billions of dollars. The implications and potential for profit were enormous — if fusion could be made to work so easily and on such a small scale, fusion power plants soon would be available in every home to provide unlimited energy.

Within days, the results were being discussed animatedly in every fusion laboratory around the world. Many fusion scientists suspected that something must be wrong — the details were sparse, but from the little that was known these results seemed to violate fundamental laws of physics. Such skepticism is the norm in scientific circles whenever unexpected results are published, but in this case, when it came from established experts in fusion, it was regarded as the malign jealousy of opposing vested interests.

Pons and Fleischmann had used the process of *electrolysis* (Box 8.1), which is well known and can be demonstrated easily in any high school laboratory. An electrolysis cell, shown schematically in Fig. 8.1, is a vessel containing a liquid (for example, water) and two metal electrodes. Passing an electric current between the electrodes ionizes the water, releasing hydrogen at one electrode and oxygen at the other. The cells used by Pons and Fleischmann contained *heavy water*, where

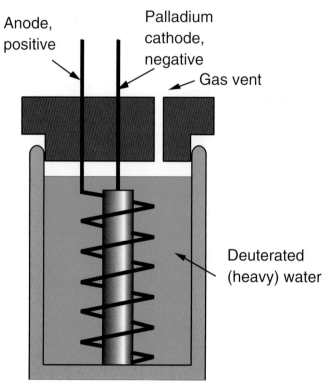

Figure 8.1 ▶ Schematic diagram of the electrolysis cells used by Pons and Fleischmann. Deuterated water (also known as *heavy water*) is contained in a glass vessel into which two metal electrodes are inserted. A current is passed through the water, and the deuterium is ionized and accumulates at the palladium cathode. It was claimed that energy was released from the cathode due to the occurrence of fusion reactions in it.

deuterium took the place of normal hydrogen, and the electrodes were of a special metal—*palladium*. It had been known for more than a century that hydrogen is absorbed in metals like palladium and can reach very high concentrations, with one or more atoms of hydrogen for every atom of metal. Pons and Fleischmann's objective was to use electrolysis to pack deuterium into the palladium to such high concentrations that atoms might get close enough for fusion.

In fact the idea was not so new—something rather similar had already been tried in Germany in the 1920s and again in Sweden in the 1930s, but these earlier claims for cold fusion had been abandoned long ago. Surprisingly, a second group, led by Steven Jones and Paul Palmer, was working on a similar topic in parallel to Pons and Fleischmann, only 30 miles away at Brigham Young University, but they were much more cautious in their claims. The competition between the two groups was one of the reasons that prompted the hurried press conference.

Section 8.1 **Fusion in a Test Tube?**

| BOX 8.1 | Electrolysis |

Electrolysis is the passing of an electric current through a liquid that contains electrically charged ions, known as an *electrolyte*. Water, being covalent, has very few ions in it, but the addition of a dilute acid or a salt such as sodium chloride provides sufficient ions to pass significant current. If sulfuric acid is used, as in the rechargeable battery of an automobile, the ions produced are H^+ and $(SO^4)^-$. The positive hydrogen ions are attracted to the negative cathode, where they pick up an electron, normally combine to form a molecule, and are released as a free gas. The negative sulfate ions go to the positive anode, where they are neutralized, but then react with water to produce more sulfuric acid and release oxygen. The sulfuric acid acts as a catalyst, and the result is the splitting up of water into hydrogen and oxygen.

It was known that if an electrode such as titanium, tantalum, or palladium, which reacts exothermically with hydrogen, is used as the cathode, the metal becomes charged with hydrogen. Instead of being released as free gas, the hydrogen enters the metal and diffuses into the lattice. Hydrides can form with concentrations of up to two hydrogen atoms per metal atom.

The electrolysis cells got hot, and the temperature increase was measured in order to compare the energy that came out with the energy put into the cell by the electric current. The measurements were delicate and great care was needed to account for everything, for there were many complicating factors. Pons and Fleischmann claimed that more energy was released than was put in — a *very* startling result, and they ascribed the additional source of energy to nuclear fusion. It was natural that there should be skepticism about this interpretation. The process of hydrogen absorption in metals had been studied for many years, and it was known that the absorbed atoms are much too far apart for fusion ever to take place. It would take much more energy than normally can be found in a metal to force the deuterium nuclei close enough together for fusion. What could provide such energy? One explanation argued that the high concentration of deuterium in the palladium might cause cracking of the metal, a known effect, and that this could produce a large electric field, which might accelerate the deuterons to high-enough energies to cause them to fuse.

It was difficult for other scientists to obtain enough information to scrutinize the claims. The results had been presented at a press conference, allowing just the barest of details to be revealed, rather than by following the recognized procedure of publishing the results in a scientific journal, which would demand more serious evidence. A detailed description of the equipment was not released because it was intended to take out a patent on the process, which naturally was thought to be very valuable. This reluctance to give details of the experiments and to respond openly to questions inevitably aroused suspicions.

Later, when Pons and Fleischmann submitted a paper describing their results to the prestigious international journal *Nature*, referees asked for more details, but the authors then withdrew the paper on the grounds that they did not have time to provide the requested information. One concern that the referees expressed was that proper control experiments had not been carried out. An obvious control would have been to compare identical cells, one with heavy water and one with ordinary water. The probability of fusion in normal hydrogen is negligible, and there should have been a clear difference between the cells if fusion really was taking place, as claimed, in deuterium. Although such a control experiment was so obvious, Pons and Fleischmann never presented any evidence to show that they had done one.

There were other discrepancies and grounds for concern. A genuine deuterium fusion reaction would produce other distinctive products that could be measured — tritium, helium, and neutrons — as well as energy. Pons and Fleischmann did not discuss the helium and tritium products at all, and initially their evidence for neutrons relied indirectly on measurements of energetic gamma rays. However, Richard Petrasso's group at the Massachusetts Institute for Technology pointed out that the gamma-ray measurements presented at the press conference had the wrong energy — a peak was shown at 2.5 MeV, but the gamma ray from deuterium fusion neutrons should be at 2.2 MeV. This small but important discrepancy in the original data was never explained, but in later presentations the peak in the spectrum was shown at the correct energy. Later, when measurements of the neutron flux were shown, the number of neutrons turned out to be very much smaller than expected and was not consistent with the amount of fusion energy that was claimed. In fact, if the neutron radiation had been commensurate with the claimed heat output, it would have killed both Fleischmann and Pons! It was difficult to measure very small neutron fluxes against the background radiation from cosmic rays and naturally occurring radioactive elements. Some experts in neutron measurements thought that there might be problems with the instruments that had been used.

However, it is not in the nature of the scientific community to ignore a result just because no explanation is immediately apparent, especially when the implications are so immense. Groups large and small the world over set out to replicate the sensational results. The equipment was relatively simple and inexpensive, and many laboratories had the necessary expertise to repeat the experiments. Although hampered by the lack of published details, intelligent guesses could be made as to what the equipment looked like, and many people started to try to see if they could get evidence of cold fusion. The measurements turned out to be rather more difficult than at first thought. The amount of energy produced was really rather small, and subtle variations in the design of the cell could upset it. Indeed, Pons and Fleischmann admitted, when pressed, that even in their own laboratory the results were not reproducible. In some cells it took a long time, often tens or hundreds of hours, before the energy anomaly was observed. This might have been explained by the time required to build up to the required density of deuterium in the metal, but in some cells nothing was observed at all. This kind of irreproducibility made it difficult to convince the world at large that there was a genuine effect. Under some conditions it could be attributed to an uncontrolled variable; possibly the results were sensitive to an unknown parameter that was not being maintained constant.

With the difficulty of the experiments and the irreproducibility of the results, some groups inevitably began to get positive results and others found negative results. An interesting sociological phenomenon occurred. Those with "positive" results thought that they had succeeded in replicating the original results and looked no further — they rushed off to publish their results. But those who got "negative" results naturally thought that they might have made a mistake or had not got the conditions quite right. They were reluctant to publish their results until they had repeated and checked the measurements. This had the effect that apparently positive results were published first — giving the impression that all who attempted the experiment had confirmed the Pons and Fleischmann results. This type of effect had occurred previously with controversial scientific data. Nobel Prize winner Irving Langmuir, an early pioneer in the area of plasma physics, had studied the effect and classified it as one of the hallmarks of what he called "pathological science".

The negative results and the mountain of evidence against cold fusion were building up steadily. In spite of a large degree of skepticism that fusion was possible under these conditions, the implications were so important that many major laboratories had decided to devote substantial resources to repeating and checking these experiments. The Atomic Energy Laboratory at Harwell in the UK made one of the most careful studies. Martin Fleischmann had been a consultant at Harwell, and he was invited to describe his results in person and to give advice on how to obtain the right conditions for energy release. But even after months of careful research, scientists at Harwell could find no evidence for cold fusion. All other mainstream laboratories reached similar conclusions. The US Energy Research Advisory Board set up a panel to investigate the claims, but no satisfactory scientific verification was forthcoming. In some cases high-energy outputs were recorded, but the necessary criteria for the occurrence of fusion were never properly fulfilled.

More than 15 years later there are still a few passionate devotees who continue to believe in cold fusion, but convincing evidence is lacking. One must conclude that this was a sad episode in science. Two reputable scientists allowed themselves to be pressured into presenting their results prematurely by the desire to get patents filed and to claim the credit for making a momentous discovery.

8.2 Bubble Fusion

In 2002 the idea of fusion in a beaker was given another brief fillip when Rusi Taleyarkhan and colleagues at the Oak Ridge National Laboratory in the US claimed to have produced nuclear fusion in *sonoluminescence* experiments. They created very small bubbles (10–100 nm in diameter) in deuterated acetone (C_3D_6O) using 14-MeV neutrons and then applied sound waves that forced the bubbles first to expand to millimeter size and then to collapse. During the collapse the vapor in the bubble heats up and produces a flash of light. It had also been speculated that nuclear fusion might be induced.

Taleyarkhan and colleagues claimed to have detected both 2.45-MeV neutrons and production of tritium, characteristic of the DD reactions, which were not observed when normal (nondeuterated) acetone was used. The result was so startling that the Oak Ridge management asked two independent researchers at the laboratory, Dan Shapira and Michael Saltmarsh, to try and reproduce the result, but they were unable to find any evidence for either the neutron emission or tritium production, even though they used a more efficient detector than had Taleyarkhan. Later work by Yuri Didenko and Kenneth Suslick at the University of Illinois, studying how the acoustic energy is shared, showed that endothermic chemical reactions during the sonic compression would make it exceedingly difficult to reach the temperatures required for nuclear fusion to take place. Bubble fusion appears to be another false trail.

8.3 Fusion with Mesons

A completely novel approach to the fusion of light elements had occurred to Charles Frank at Bristol University, UK, in 1947. A group at Bristol, led by Cecil Powell, had been studying cosmic rays using balloons to carry photographic plates high up into the stratosphere. Cosmic rays entering the Earth's atmosphere from outer space were recorded as tracks on these plates, leading to the discovery of many new atomic particles, including what is now called the *mu meson*, or *muon*. The muon is 207 times heavier than an electron, and there are three versions, with positive, negative, or neutral charge. Although discovered originally in cosmic rays, muons can now be produced in particle accelerators.

A negative muon behaves rather like a very heavy electron, and Charles Frank realized that it would be possible to substitute a negative muon for the electron in a hydrogen or deuterium atom. According to the rules of quantum mechanics, the muon would have an orbit around the nucleus with a radius inversely proportional to its mass. The *muonic atom*, as it is now called, would be only 1/207 the size of a normal hydrogen atom. This means that such a muonic deuterium atom could approach very close to another muonic deuterium or tritium atom and have the possibility of fusing and releasing energy. An equivalent way of looking at the situation is to say that the negative charge of the muon, in close orbit around the nucleus, shields the positive charge, allowing two nuclei to get close enough to each other to fuse (Fig. 8.2).

There are two problems to any practical application of this idea. The first is that muons are unstable, with a lifetime of only 2.2 microseconds. The second is that it takes more energy to produce a single muon in an accelerator than the energy released by a single DT fusion reaction — about 1000 times more in fact. However, all was not yet lost. Frank realized that there was the possibility that the

Section 8.3 Fusion with Mesons

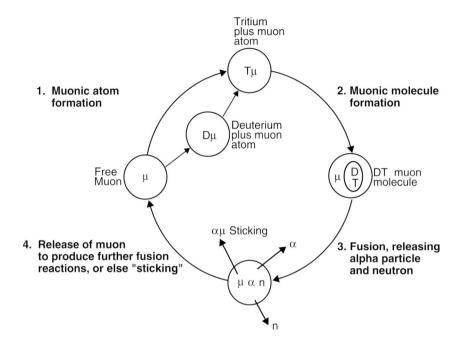

Figure 8.2 ▶ Schematic of the four stages of muon-catalyzed fusion. (1) The negative muon is captured by positive deuterium or tritium nuclei to form a muonic atom. The process tends to form muonic tritium atoms because they are slightly more stable than deuterium atoms. (2) The muonic tritium atom then combines with an ordinary deuterium molecule to form a complex muonic molecule. (3) The small size of the muon orbit partially screens the positive charges on the nuclei, allowing them to get close enough to each other to undergo a fusion reaction, forming an alpha particle and a neutron. (4) The alpha particle flies off, leaving the muon to repeat the process. But in some cases it sticks to the muon, preventing further reactions.

muon would be released after the DT fusion reaction, since the alpha particle would go off with high velocity, freeing the muon to initiate another fusion reaction. The muon would act rather like a catalyst, and the process is known as *muon-catalyzed fusion*. Obviously the key to making this viable as a source of energy is that each muon must be able to catalyze at least 1000 fusion events in its short lifetime in order to generate more energy than has to be invested in making it.

In the Soviet Union, Andrei Sakharov took up Frank's idea. But his calculations of the rate of muon-catalyzed fusion at first appeared to indicate that the process was too slow to enable a net gain in energy within the lifetime of the muon. Meanwhile the first observation of muonic fusion came in 1956 as a by-product of an accelerator experiment by Luis Alverez in Berkeley, California. Dedicated muonic fusion experiments were subsequently conducted in the Soviet Union, first with DD and later with DT, demonstrating that the catalytic process works in principle. At around the same time, two Soviet physicists, Sergei Gerstein and

Leonid Ponomarev, realized that the original theory was inadequate and that a muonic DT molecule could form much more rapidly than had previously been realized. There was thus the possibility that even in the short life of the muon it would be possible to have many fusion reactions.

This work led to renewed interest in the subject, and new experiments were undertaken to determine just how many fusion reactions per muon could be achieved. Experiments led by Steven Jones (later to be involved in the cold fusion work at Brigham Young University) at Los Alamos Laboratory in New Mexico used a gold-lined stainless steel vessel containing deuterium and tritium that could be controlled over a wide range of temperature and pressure. The theory indicated that the molecule formation would be faster at high pressure, and results published in 1985 showed that increasing the pressure of the DT mixture increased the number of fusions per muon to about 150. However, this was still very short of the number required for net production of energy. The parallel with "hot" fusion of the period, whether by magnetic confinement or by inertial confinement, was striking. It was possible to release fusion energy, but it was still necessary to put in much more energy than it was possible to get out.

The main problem is a phenomenon called *sticking*. Each fusion reaction creates a positively charged alpha particle that has the possibility of attracting and trapping the negative muon so that it is no longer available for fusion. For example, if the sticking probability were 1%, the maximum number of muon-catalyzed fusions would be 100, too low to make the process economical. Recent experimental work has concentrated on trying to measure the sticking probability and to see if it can be reduced.

Some of the most recent experiments have been done by a joint UK–Japanese collaboration, headed by Kanetada Nagamine. The experiments are carried out at the Rutherford Appleton Laboratory in the UK using the ISIS accelerator, the most intense source of muons presently available in the world. Nagamine has measured a sticking probability of 0.4% that allows about 250 fusion events for each muon. Ideas for decreasing the sticking probability are being studied. However, the realization seems to be gradually developing that this value of the sticking probability is a fundamental limit. It seems an unfortunate fact of nature that it is possible to get so tantalizingly close to energy breakeven by this method without being able to achieve it.

▶ Chapter 9

Tokamaks

9.1 The Basics

Chapter 5 described the basic principles of magnetic confinement, but the story stopped in 1968 at the international fusion conference in Novosibirsk (Fig. 9.1), just at the time when tokamaks had overtaken toroidal pinches and stellarators. The only cause to doubt the tokamak results was that the temperature of the tokamak plasma had been measured rather indirectly. To put the issue beyond all doubt, a team of British scientists was invited to go to Moscow the following year to measure the temperature with a new technique based on lasers (see Box 9.4). When these results were published, even the most skeptical fusion researchers were convinced. The leading American fusion laboratory, at Princeton, moved quickly, converting its biggest stellarator into a tokamak. This was followed by the construction of many new tokamak experiments in the fusion laboratories of the US, Europe, Japan, and, of course, the Soviet Union. Thereafter tokamaks set the pace for magnetic-confinement fusion research.

The basic features of a tokamak are shown in Fig. 9.2. There are two main magnetic fields. One of these, known as the *toroidal* field, is produced by a copper coil wound in the shape of a torus. The second magnetic field, the *poloidal* field, is generated by an electric current that flows in the plasma. This current is induced by a transformer action similar to that described for the pinch system in Chapter 5. The two magnetic fields combine to create a composite magnetic field that twists around the torus in a gentle helix. Superficially a tokamak resembles a toroidal pinch, which also has two similar fields. The main difference is in the relative strength of the two components. In a pinch, the poloidal field is much stronger than the toroidal field, so the helix is very tightly twisted, but in a tokamak it is the other way round — the toroidal field produced by the external coils is typically about 10 times stronger than the poloidal field due to the plasma current.

Figure 9.1 ▶ A group of Soviet and British scientists during the Novosibirsk conference in 1968, at which it was agreed that a team from the UK would go to Moscow to confirm the high plasma temperature claimed for the tokamak, using the new technique of scattering of laser light. From left to right: Bas Pease, Yuri Lukianov, John Adams, Lev Artisimovich, Hugh Bodin, Mikhail Romanovski, and Nicol Peacock.

The main effect of the toroidal field is to stabilize the plasma. The most serious instabilities have to bend or stretch this magnetic field, which acts rather like the steel bars in reinforced concrete (although plasma is more like jelly than concrete). The fact that a tokamak has stronger "reinforcing bars" than a pinch has the effect of making it much more stable. Stellarators also have a strong toroidal field and share the inherent stability of the tokamak. However, because they have no plasma current, they require additional complex external coils to generate the poloidal magnetic field. These coils are expensive to build and must be aligned with great accuracy. So one of the reasons tokamaks overtook stellarators and stayed out in front is simply that they are easier and cheaper to build.

9.2 Instabilities

Trying to contain hot plasma within a magnetic field is rather like trying to balance a stick on one's finger or position a ball on top of a hill — any small displacement grows with increasing speed. Plasma is inherently unstable and always tries to escape from the magnetic field. Some types of instability cause the sudden loss of the whole plasma; others persist and reduce the energy confinement time. Although these instabilities make it difficult to reach and maintain the

Section 9.2 **Instabilities**

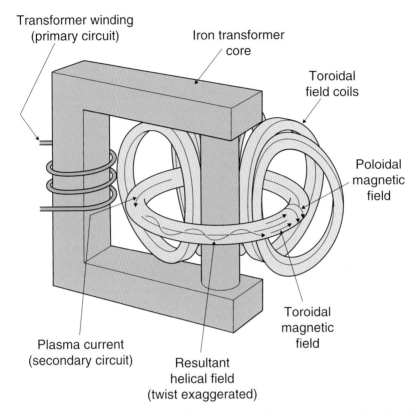

Figure 9.2 ▶ A schematic view of a tokamak showing how the current is induced in the plasma by a primary transformer winding. The magnetic fields, due to the external coils and the current flowing in the plasma, combine to produce a helical magnetic field.

conditions for fusion, it is important to stress that there is no danger, in the sense of causing explosions or damage to the environment. Plasma that escapes from the magnetic field cools quickly as soon as it touches a solid surface.

Even with the strong toroidal magnetic field, tokamak plasmas become unstable if either the plasma current or the plasma density is increased above a critical value. The plasma extinguishes itself in a sudden and dramatic manner, known as a *disruption* (Box 9.1). Disruptions can be avoided to some extent by careful selection of the operating conditions. The maximum current and density depend on the tokamak's physical dimensions and on the strength of the toroidal magnetic field (Box 10.1, p. **??**). Disruptions occur when the magnetic field at the plasma edge is twisted too tightly. Increasing the plasma current increases the poloidal magnetic field, enhances the twist, and makes the plasma unstable. Increasing the toroidal field straightens out the twist and improves stability. The effect of plasma density is more complicated. In simple terms, increasing density causes

BOX 9.1 Disruptions

A major disruption in a tokamak is a dramatic event in which the plasma current abruptly terminates and confinement is lost. It is preceded by a well-defined sequence of events with four main phases. First there are changes in the underlying tokamak conditions — usually an increase in the current or density, but not always clearly identified. When these changes reach some critical point, the second phase starts with the increase of magnetic fluctuations in the plasma, whose growth time is typically of the order of 10 ms. The sequence then passes a second critical point and enters the third phase, where events move on a much faster time scale — typically of the order of 1 ms. The confinement deteriorates and the central temperature collapses. The plasma current profile flattens, and the change in inductance gives rise to a characteristic negative voltage spike — typically 10–100 times larger than the normal resistive loop voltage. Finally comes the current quench — the plasma current decays to zero at rates that can exceed 100 MA/s. Disruptions cause large forces on the vacuum vessel, and these forces increase with machine size. In big tokamaks forces of several hundred tons have been measured, and a power plant would have to withstand forces at least an order of magnitude higher.

One of the factors determining when a disruption occurs is the amount by which the magnetic field lines are twisted. The twist is measured by the parameter q, known as the *safety factor*, which is the number of times that a magnetic field line passes the long way around the torus before it returns to its starting point in the poloidal plane — large q indicates a gentle twist and small q a tight twist. Usually the plasma becomes unstable when the parameter is $q < 3$ at the plasma edge. (See Box 10.1, p. ??, for a more detailed discussion.)

Attempts to control disruptions have met with only limited success. Applying feedback control to the initial instability has delayed the onset of the third phase, but only marginally. Power-plant designers have proposed that disruptions should be avoided at all costs, but this seems unrealistic. A possible remedy is the injection of a so-called "killer pellet" to cool the plasma by radiation before the onset of the current quench.

---

the plasma edge to become cooler because more energy is lost by radiation, as discussed later. Because hot plasma is a much better electrical conductor than cool plasma, the cooler edge squeezes the current into the core, increasing the twist of the magnetic field by reducing the effective size of the plasma.

Some tokamak instabilities are subtler and do not completely destroy the plasma but result in loss of energy. For example, the temperature in the center of a tokamak usually follows a regular cycle of slow rises and rapid falls. A graph of temperature against time looks a bit like the edge of a saw blade and gives the name *sawteeth* to this instability (Box 9.2). Sawteeth are linked to the twist of the magnetic field in the center of the plasma. Very small-scale instabilities seem to appear all the time in tokamaks. Too small to be easily seen, they are held responsible for reducing plasma confinement (Box 10.3, p. ??). Tokamak instabilities are a very complex subject, and they are still far from being fully understood; they are similar in many ways to the instabilities in the weather, with which they have much in common, because both are controlled by the flow of fluids and show turbulent behavior.

BOX 9.2 **Sawteeth**

The central temperature often varies in a periodic way, increasing in a linear ramp to a peak value, dropping abruptly, and then starting to increase again (Fig. 9.3). This feature is described as *sawteeth*, for obvious reasons.

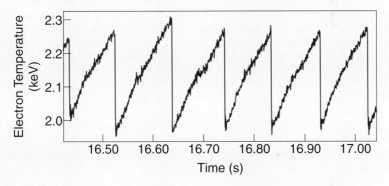

Figure 9.3 ▶ The variations of central temperature in a tokamak plasma, illustrating typical sawteeth.

From observations at different radii, a flattening of the central temperature profile is observed when the temperature drops. There is also an increase in the temperature further out indicating a sudden outflow of energy. A model proposed by Boris Kadomtsev from the Kurchatov Institute in Moscow was initially rather successful in explaining the behavior in terms of a magnetic *island*, which grows and tangles up the field lines. Eventually the tangled field sorts itself out by *reconnecting* — in effect the field lines snap and rejoin in a more ordered way. The reconnection allows plasma in the hot central region to mix with cooler plasma. However, recent measurements of sawteeth reveal many details that are in disagreement with the Kadomtsev model, and, although attempts have been made to improve the theory, no model of sawteeth instability has yet been generally accepted.

It used to be assumed that sawteeth are triggered when the safety factor q reached unity in the center of the plasma (theory shows this is unstable). When techniques were developed to measure q directly inside a tokamak, it came as a surprise to find $q < 1$.

9.3 Diagnosing the Plasma

The first stage of operation is getting the tokamak working (Box 9.3). The second stage is *diagnosing* the plasma — measuring its properties (Box 9.4). This is difficult because the plasma is so hot that it is not possible to put any solid probe deep

---

BOX 9.3 Operating a Tokamak

The start-up phase of any new experiment is always exciting. There are many different components and systems, and all have to be working simultaneously before a plasma can be produced and heated. Although experience from the operation of other tokamaks is valuable, the operation of each one has its own individual characteristics. An analogy is that of learning to drive an automobile of the 1900s era. The driver knew roughly the right things to do, but it required trial and error to learn each car's idiosyncrasies, to learn how to change gear and make it operate in the smoothest and most efficient manner. In a similar way the tokamak operation has to be slowly optimized. It is known that there are limits to the plasma current, pressure, and density, above and below which the plasma will not be stable (for more detail see Box 10.1, p. ??). Where will these limits be in the new tokamak? Will the confinement time and the temperature be as predicted?

After the power supplies and control electronics have been checked, the first thing to do is to pump air out of the vacuum vessel and clean it to reduce the impurities. Then the magnetic fields are turned on, deuterium gas is introduced, and a voltage is applied. This voltage causes the gas to break down electrically, become ionized, and form a plasma. As it becomes electrically conducting, a current begins to flow. The plasma current and the length of the current pulse are determined by the capacity of the power supplies. After each plasma pulse the current falls, the plasma cools down, and the power supplies are recharged for the next pulse. The repetition rate for experiments is typically one pulse every 15 minutes. Once the plasma is being produced reliably, work starts on the detailed measurements of the various parameters of the plasma. Little by little, as confidence in the machine operation increases, the magnetic fields and the current in the plasma are raised. As the parameters increase it is necessary to check all the mechanical and electrical systems to see that they are not being overstressed. It takes some months, even years, of operation before the operators find the optimum plasma conditions. This phase is common to many physics experiments and is analogous to the optimization of accelerator operation in high-energy physics. When all the components are operating reliably, the physics starts.

After typically 3–6 months of operation, the experiments stop so that maintenance can be carried out. Modifications and improvements are made in light of the previous operating experience. These *shutdowns* can last from a few weeks to over a year. Because the machines are operating near the limits of endurance of materials and because they are built as experiments rather than production machines, improved operation has to be acquired gradually.

Section 9.3 **Diagnosing the Plasma** 101

| BOX 9.4 | **Temperature Measurement** |

It is important in fusion experiments to be able to measure the properties of the plasma. These include temperature, density, impurity concentration, and particle and energy confinement times. Because of the high temperatures involved, measurements have to be made remotely, usually by observing radiation from the plasma. Electron temperatures are measured by firing a laser beam into the plasma and detecting the light scattered by the electrons. The scattered light is Doppler shifted in frequency due to the temperature of the electrons. By viewing different positions along the length of the laser path, a radial profile of temperature can be measured (Fig. 9.4).

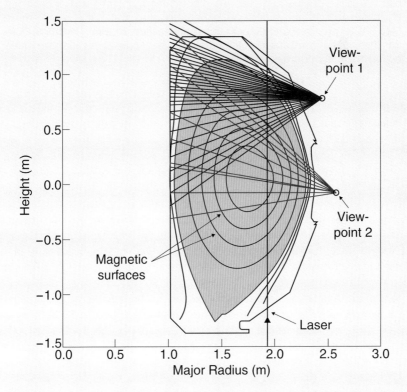

Figure 9.4 ▶   The Thomson scattering diagnostic for the DIII-D tokamak in San Diego. The laser beam passes through the plasma along a vertical line. Up to 40 detector channels receive light scattered via the two viewpoints.

It is usually justified to assume that the electrons have a Maxwellian distribution of energies, and it is thus sufficient to measure the photon flux at a few discrete wavelengths to obtain the temperature. The local electron density can also be

> **BOX 9.4** (continued)
>
> obtained from the total number of photons scattered from a specific volume. The number of photons scattered is generally small, so a high-energy laser (ruby or neodymium) pulsed for a short interval is employed. The laser may be pulsed repetitively to obtain the time evolution of the temperature. Electron temperature can also be measured from *electron cyclotron radiation*, emitted due to the acceleration of the electrons as they follow circular orbits in the magnetic field.
>
> Impurity ion temperatures can be measured directly using a high-resolution spectrometer to observe the Doppler broadening of a suitable spectral line. Because impurities have different charge states in different regions of the plasma, it is possible to measure temperatures in different radial positions. Hydrogen ions in the plasma core cannot be measured in this way because they are fully ionized and do not emit spectral lines. Normally hydrogen ions and the impurities are at similar temperatures, and some correction can usually be made for any discrepancy between them.

into it. Much can be learned by analyzing the radiation emitted from the plasma: X-rays, light, radio emission, and neutral particles. More information is obtained by injecting laser beams, microwaves, and beams of neutral atoms into the plasma and observing their attenuation or the light scattered or emitted from them. Many of these measurements involve sophisticated computer-controlled instruments and analysis. From the data obtained it is possible to calculate the density and the temperature of the electrons and ions, the energy confinement time, and many other properties. Measurements have to be made at many different positions because of the variation in the plasma properties, from the hot and dense core to the relatively cool edge. Another type of investigation is studying the instabilities in the plasma to determine the conditions that induce the instabilities and the effect that the instabilities have on the loss of energy.

When the plasma has been properly diagnosed, the next stage is to compare its properties with the predictions of various theoretical models. In this way it is possible to get a better understanding of why the plasma behaves in the way it does. Only with this understanding is it possible to see how to improve the plasma and to get closer to the conditions required for a fusion power plant.

9.4 Impurities

Tokamak plasmas are normally contained in a stainless steel toroidal vacuum vessel that nests inside the toroidal magnetic coils. All the air inside the vessel is evacuated, and it is filled with low-pressure deuterium gas that is ionized to create plasma. Although ionized from pure deuterium gas, plasma can quickly become contaminated with other elements. These are known as *impurities*, and

BOX 9.5 | Sources of Impurities

In a burning plasma, the fusion process is an internal source of helium ash. Other impurities are released from the material surfaces surrounding the plasma by a variety of processes. There are impurities, particularly carbon and oxygen, that are entrained in the bulk of the metal during manufacture and migrate to the surface. These surface contaminants are released by radiation from the plasma or as a result of sputtering, arcing, and evaporation. *Sputtering* is a process in which energetic ions or neutrals knock atoms, including metal atoms, from a surface by momentum transfer. *Arcing* is driven by the voltage difference between the plasma and the surface. *Evaporation* occurs when the power load is sufficient to heat surfaces to temperatures near their melting point—this is often localized in hot spots on prominent edges exposed to plasma. All three mechanisms are important at surfaces that are subject to direct plasma contact, such as the limiter or divertor, but generally the walls are shielded from charged particles and are subject only to sputtering by charge-exchange neutrals.

---

their main source is the interaction between the plasma and the material surfaces (Box 9.5).

Layers of oxygen and carbon atoms always cover surfaces in vacuum, even under the most stringent vacuum conditions. In fact a perfectly clean surface is impossible to maintain—a clean surface is so "sticky" that it quickly becomes coated again. Particles and radiation from the plasma bombard the wall. This bombardment dislodges some of the oxygen and carbon atoms sticking to the surface and even knocks metal atoms out of the surface itself (Box 9.5). These impurity atoms enter the plasma, where they become ionized and trapped by the magnetic fields. Impurity ions, like oxygen and carbon, with not too many electrons radiate energy and cool the plasma most strongly near the plasma edge (Box 9.6). Too much edge cooling makes the plasma unstable, and it disrupts when the density is raised. In order to reach high plasma density, the concentrations of oxygen and carbon impurities have to be reduced to less than about 1%. Metal ions, especially those with many electrons, like tungsten, are responsible for much of the energy radiated from the core. Too much cooling in the core makes it difficult to reach ignition temperatures.

The effects of impurities are reduced by careful design and choice of materials for the internal surfaces of the tokamak. The internal surfaces are carefully cleaned; the toroidal vacuum chamber is heated up to several hundred °C and conditioned for many hours with relatively low-temperature plasmas before a tokamak operates with high-temperature fusion plasmas. Soviet physicists had a neat term for this process; they referred to "training" their tokamaks. However, even with the best training there were still impurities from the walls of the vacuum chamber in which the plasma was confined.

A very important feature of tokamaks is that there must be a well-defined point of contact between plasma and wall that can withstand high heat loads. This is

> **BOX 9.6** **Impurity Radiation**
>
> Impurities in tokamak plasmas introduce a variety of problems. The most immediate effect is the radiated power loss. A convenient parameter to characterize the impurity content is the *effective ion charge*, $Z_{eff} = \Sigma_i n_i Z_i^2 / n_e$, where the summation is taken over all the ionization states of all the ion species.
>
> Line radiation from electronic transitions in partially-ionized impurity ions is the most important cooling mechanism, but it ceases at the temperature where the atoms are fully ionized and lose all their electrons. Hence, radiative cooling is particularly significant in the start-up phase, when the plasma is cold, and also at the plasma edge. *Low-Z* impurities, such as carbon and oxygen, will lose all their electrons in the hot plasma core. But to reach ignition, one must overcome the radiation peak, which for carbon is around 10 eV. Next, the plasma must burn through the radiation barriers presented by any *medium-Z* impurities, such as iron and nickel, and *high-Z* impurities, such as molybdenum and tungsten, around 100 eV and 1 keV, respectively. A DT plasma with as little as 0.1% of tungsten would radiate so much power that it would be impossible to reach ignition.
>
> From the point of view of radiation cooling, much larger concentrations of carbon, oxygen, and helium ash could be tolerated in a fusion power plant, but then the problem of fuel dilution arises. An impurity ion produces many electrons. In view of the operating limits on electron density and plasma pressure (see Box 10.1, p. ??), this has the effect of displacing fuel ions. For example, at given electron density, each fully ionized carbon ion replaces six fuel ions, so a 10% concentration of fully ionized carbon in the plasma core would reduce the fusion power to less than one-half of the value in a pure plasma.

called a *limiter* because it limits the size of the plasma. Limiters in the early tokamaks were usually made of metals, like tungsten and molybdenum, that can withstand high temperatures. However, because these metals cause serious problems if they get into the plasma, many tokamaks now use materials like carbon and beryllium that have less impact as impurities. But carbon and beryllium bring other problems, and there is no ideal choice of wall material for a fusion power plant — this is one of the issues still to be resolved.

Suggestions had been put forward as early as the 1950s by Lyman Spitzer at Princeton of a method of mitigating this wall interaction in his stellarator designs. He proposed a magnetic-confinement system where the magnetic field lines at the edge of the plasma are deliberately *diverted*, or led away, into a separate chamber, where they can interact with the wall. This system is called a *divertor*.

Impurities produced in the divertor are inhibited from returning to the main chamber by restricting the size of the holes between the two chambers and because the flow of the main plasma from the main chamber tends to sweep impurities back into the divertor, rather like a swimmer in a strong current. When the stellarator fell out of favor, this idea was applied to the tokamak, with considerable success (Figs. 9.5 and 9.6).

Section 9.4 **Impurities**

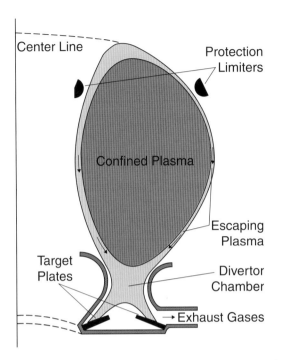

Figure 9.5 ▶ An example is shown of a type of tokamak divertor at the bottom of the machine. The poloidal magnetic field generated by the plasma current is distorted by extra coils (not shown), which produce a magnetic field that is locally in the opposite direction to the poloidal field. This allows the plasma escaping from the confined region to stream down into the divertor chamber and to interact with a surface (the divertor target) that is remote from the confined hot plasma.

Figure 9.6 ▶ Plasma light emission from the divertor region in the ASEDEX-U tokamak in Munich, Germany. The bright rings of light are the slots where the plasma makes the strongest contact within the director.

The divertor is unfortunately a significant complication to the tokamak design, making it more difficult to build and more expensive and, for a given physical size of magnetic field coil, reducing the volume available for the confined plasma. The arguments for and against a divertor are therefore somewhat finely balanced. However, a strong additional argument in favor of the divertor is the role it can play in removing the helium produced by fusion reactions. If the helium is not removed it will accumulate and dilute the deuterium and tritium fuel so that the fusion output is reduced and the conditions for ignition are lost. It can be looked on as the "ash" from a fusion power plant. A divertor helps to concentrate the helium so that it can be pumped away and removed.

9.5 Heating the Plasma

The electric current in the plasma has two functions. As well as generating the poloidal magnetic field, it heats the plasma, in the same way that an electric current flowing through a metal wire heats it up. This heating effect, known as *ohmic-* or *joule-heating*, was very efficient in early tokamaks and was one of the reasons for their success. However, as the plasma gets hotter, its electrical resistance falls and the current becomes less effective at heating. The maximum temperature that can be reached with ohmic heating is typically less than 50 million degrees — very hot by most standards but still some way to go on the scale required for fusion.

The most obvious solution would appear to be to increase the plasma current. However, increasing it too much will cause the plasma to disrupt (Box 9.1), unless the toroidal magnetic field is increased proportionally (Box 10.1, p. ??). The forces on the magnetic coils increase rapidly with field and set the upper limit to the strength of the toroidal field. Some tokamaks have pursued this approach to very high magnetic fields. A spin-off benefit is that the maximum plasma density and the energy confinement time increase with magnetic field. Early success gave rise to optimism that the conditions for ignition could be reached in this way — but there are severe engineering difficulties.

The need for additional plasma heating to give control of plasma temperature independent of the current has led to the development of two main heating techniques. The first technique, known as *neutral-beam heating*, uses powerful beams of energetic neutral deuterium atoms injected into the plasma (Box 9.7).

The second plasma-heating technique uses radio waves (Box 9.8), rather like a microwave oven, where food is heated by absorbing energy from a microwave source. The plasma version requires a bit more sophistication to select a radio frequency at which the plasma will absorb energy. The most important frequency ranges are linked to the natural resonance frequencies of ions and electrons in the toroidal magnetic field. The ion resonance frequency is in a range similar to that used for radio and television transmitters, and the electron resonance frequency is in a range close to that used for radar. Using a combination of neutral beam and radio wave heating provides a flexible arrangement and allows some control over the temperature of both the electrons and the ions in the plasma.

BOX 9.7 Production of Neutral Heating Beams

The neutral beams used for plasma heating start life outside the tokamak as deuterium ions that are accelerated to high energies by passing them through a series of high-voltage grids (Fig. 9.7). A beam of energetic ions cannot be injected directly into a tokamak because it would be deflected by the magnetic fields. The ion beam is therefore converted into neutral atoms by passing it through deuterium gas, where the ions pick up an electron from the gas molecules. The beams of neutral atoms are not deflected by magnetic fields and can penetrate deep inside the plasma, where they become reionized. After being ionized, the beams are trapped inside the magnetic fields and heat the plasma by collisions.

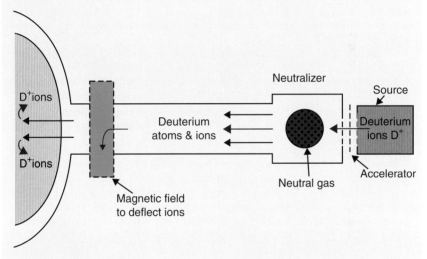

Figure 9.7 ▶ Schematic of a neutral beam injection system for plasma heating.

The energy required for the neutral beams has to match the size of the tokamak plasma and its density. The typical beam energy for present-day tokamaks is about 120 keV, but the neutral beams for the next generation of tokamaks and for fusion power plants require much higher energy, up to 1 MeV. At these higher energies the process of neutralizing a positive ion beam becomes very inefficient, and a new type of system that accelerates negative rather than positive ions needs to be developed.

Much work has been done over many years developing high-current sources in order to inject the maximum power into tokamaks. There is a theoretical limit to the amount of ion current that may be extracted through a single hole in the accelerating structure, and grids with multiple holes, up to more than 200, are used. One of the technical difficulties is removing the waste heat in this system. Currents of about 60 amps of ions are routinely generated at 120 keV, but the neutralization efficiency is only about 30%, leading to an injected power of just over 2 MW of neutrals per injector.

> **BOX 9.8**   Radiofrequency Heating
>
> The highest radio frequency used for plasma heating is determined by the *electron cyclotron resonance*, $\omega \approx \omega_{ce}$, which depends only on the toroidal magnetic field. The resonance, at 28 GHz/T (gigahertz per tesla), falls in the frequency range 60–120 GHz for most experiments, but frequencies up to 200 GHz will be required for a power plant. The free-space wavelength is in the millimeter waveband. Waves outside the plasma can be propagated in metal waveguides, and the launching antennas can be placed well back from the plasma edge. Electron cyclotron waves heat the electrons, which in turn transfer energy to the ions by collisions. The method may be used for both global and localized heating and has been applied to control sawteeth and disruptions. The application of electron cyclotron heating has been hindered by a lack of powerful, reliable high-frequency sources. But this is now being overcome, and it is thought that these techniques will play an increasingly important role in fusion.
>
> Lower in frequency, the *ion cyclotron resonance*, $\omega \approx \omega_{ci}$, depends on the *charge-to-mass ratio* of the ion (Z/A) and on the toroidal magnetic field. The fundamental resonance lies at 15.2 (Z/A) MHz/T. The technology of power generation and transmission is well developed because the frequency range of interest for fusion, 40–70 MHz, is the same as that widely used for commercial radio broadcasting. In a tokamak, the toroidal field falls off as $1/R$, so the heating is localized at a particular value of *radius*, determined by the frequency. The physics of the heating process is quite complicated and requires either a plasma with two ion species (for example, a concentration of a few percent hydrogen minority in a deuterium plasma) or working at the second harmonic resonance $\omega \approx 2\omega_{ci}$. Antennas must be placed very close to the plasma edge for efficient coupling because the waves cannot propagate below a certain critical plasma density (typically about 2×10^{18} m^{-3}). Plasma interaction with these antennas, aggravated by the localized intense radio frequency fields, can lead to a serious problem with impurity influx. In spite of its name, ion cyclotron heating is usually more effective in heating the plasma electrons than in heating the plasma ions. A combination of neutral beam heating and ion cyclotron heating allows some degree of control of both ion and electron temperatures.
>
> There is a third resonance frequency, midway between the ion and electron cyclotron frequencies, known as the *lower hybrid resonance*. It falls in the range of 1–8 GHz, corresponding to free-space wavelengths in the microwave region of the spectrum. This has proved less effective for plasma heating but is used to drive currents in the plasma.

These heating techniques were introduced to tokamaks in the 1970s and soon plasma temperatures began to rise toward the range required for nuclear fusion. At first it seemed that the fusion goal was close to hand, but the temperature rise was less than expected. It was already known that energy loss from a tokamak plasma was faster than expected, but careful studies now showed that the loss increased further as the heating power was increased. In other words the energy

Section 9.5 **Heating the Plasma**

| BOX 9.9 | L- and H-Modes |

During neutral beam heating of the ASDEX tokamak in 1982, Fritz Wagner and his colleagues found that under certain conditions there was an abrupt transition in the plasma characteristics and the energy and particle confinement improved. The behavior was unexpected, but subsequently this type of transition was observed on other tokamaks with divertors. It has also been observed, but with more difficulty, in tokamaks with limiters. The state of the plasma after the transition has been called the *H-mode* (for high confinement), to distinguish it from the usual *L-mode* (for low confinement). The improvement in energy confinement is typically a factor of 2.

The onset of the H-mode requires the heating power to be above a certain threshold. The transition is first observed at the edge of the plasma, where there is a rapid increase in both the edge density and temperature, resulting in steep edge gradients. It can be simply thought of in terms of an edge transport barrier. The edge and central temperatures and densities rise, but the new equilibrium is not easily controlled. Usually the edge region becomes unstable (Fig. 9.8) with periodic bursts of plasma loss known as *edge-localized modes (ELMs)*.

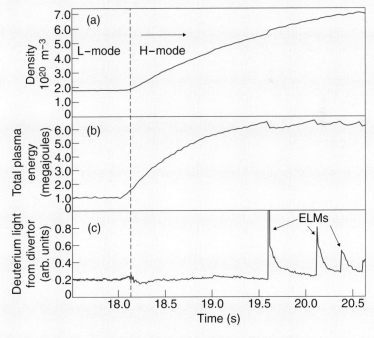

Figure 9.8 ▶   Illustration of typical behavior of the transition from L- to H-mode showing: (a) the density, (b) the plasma energy, and (c) the light emitted by deuterium atoms in the divertor, showing the spikes characteristic of ELMs.

confinement time deteriorates as the heating is applied. An analogy is a house with central heating where the windows open more and more as the heating is turned up. The room can still be heated, but it takes more energy than if the windows stay closed. Plasma temperatures in the range required for fusion can be reached, but it takes much more heating than originally expected.

This was bad news indeed, but there was a ray of hope when some experiments started to show better confinement than others. A new operating regime with high confinement was discovered in the ASDEX tokamak in Germany by combining a divertor with neutral beam heating; it became known as the *H-mode* (Box 9.9).

▶ Chapter 10

From T3 to ITER

10.1 The Big Tokamaks

The success of the newly-built tokamaks in the early 1970s encouraged fusion scientists to develop ambitious forward-looking plans. In the wake of the oil crisis of 1973, alternative energy sources enjoyed renewed popular and political support and the prospects for funding fusion research improved. However, extrapolating to the plasma conditions for a fusion power plant from the relatively small tokamaks then in operation was a very uncertain procedure. Plasma temperatures had reached less than one-tenth of the value required for ignition, and energy confinement times had even further to go. It was clear that powerful heating would be required to reach the temperature range for ignition, but heating techniques (see Chapter 9) were still being developed, and it was too early to say how energy confinement would scale.

Although theoretical understanding of the energy and particle loss processes was lacking, experimental data from the early tokamaks showed that both plasma temperature and energy confinement time appeared to increase with plasma current. Though there was much scatter in the data, the projections available at that time suggested that a tokamak with a current of about 3 million amps (3 MA) might reach the point known as *breakeven*, where the fusion energy output matches the required heating input. This would still fall short of *ignition* (the condition where heating by alpha particles is sufficient to sustain the plasma without any external heating) but would be a crucial step toward making the task of designing a fusion power plant more certain.

The French TFR and Russian T4 tokamaks were the most powerful in operation in the early 1970s, with maximum plasma currents of about 400,000 amps. The Americans were building a 1-MA tokamak, known as the *Princeton Large Torus (PLT)*, due to start operating in 1975. The Europeans decided on a bold step and in 1973 started to design a 3-MA tokamak — the Joint European

Torus (JET). Too big and expensive for a single fusion laboratory, this became a collaborative venture, with most of the funding provided by the European Community. The design progressed rapidly, but political decisions came slowly. Both Britain and Germany wanted to host JET. The ministers most directly concerned were Anthony Wedgewood Benn and Hans Matthöfer, and neither would give way. The crisis came in October 1977 as the design team was beginning to disintegrate, and the resolution came about in a rather surprising way. On October 17 there was a spectacular rescue at Mogadishu, Somalia, of a Lufthansa plane that had been hijacked by the Baader-Meinhof gang. The UK had supplied special grenades used to stun the terrorists. In the general atmosphere of goodwill following the rescue it appears that Chancellor Helmut Schmidt and Prime Minister James Callaghan came to an understanding. The decision to build JET at Culham, just south of Oxford in the UK, was finally made at a meeting of the European Community research ministers on October 25.

The initial European mandate to build a 3-MA tokamak was extended to 4.8 MA during the design phase. In fact, JET (Fig. 10.1) had been well designed by the team, led by Paul Rebut (Fig. 10.2), and was able to reach 7 MA. The contrast with earlier fusion experiments, many of which had failed to reach their design goals, was striking. Water-cooled copper coils provided JET's

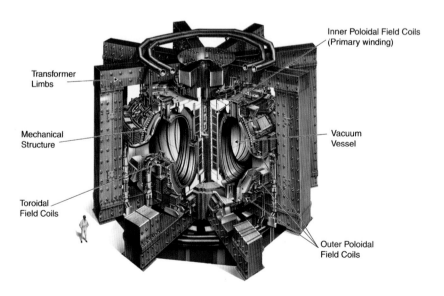

Figure 10.1 ▶ A cutaway model of the JET tokamak, showing the iron transformer, the magnetic field coils, and the vacuum vessel in which the plasma is formed. Built as the flagship experiment of the European Community fusion program, JET remains the largest tokamak in the world.

Section 10.1 **The Big Tokamaks**

Figure 10.2 ▶ Paul-Henri Rebut, the leader of the JET design team from 1973 to 1978, deputy director of JET from 1978 to 1987, director from 1987 to 1992, and director of ITER from 1992 to 1994.

magnetic field. These were "D-shaped," in order to minimize the effects of electromechanical forces. The elongated shape allowed plasma with bigger volume and increased current as compared to a circular cross section, and later it allowed the installation of a divertor (Fig. 10.3).

Peak electrical power for magnets, heating, and other systems requires 900 MW, a substantial fraction of the output of a big power plant. Taking such a large surge of electrical power directly from the national electricity grid would have side effects for other customers, so two large flywheel generators boost the power for JET. These generators, with rotors weighing 775 tons, are run up to speed between tokamak pulses using a small motor to store energy in the rotating mass. When the rotor windings are energized, the rotational energy is converted to electrical energy. Delivering up to 2600 MJ of energy with a peak power of 400 MW, each generator slows down to half speed in a few seconds; after that the power supplies connected directly to the grid take over.

Europe's decision to design a big tokamak provoked a quick response from the Americans. Though slightly later in starting the design of their Tokamak Fusion

Figure 10.3 ▶ A wide-angle view (taken in 1996) inside JET. Most of the inner surfaces are covered with carbon tiles. The divertor can be seen as the annular slots at the bottom. The man wearing protective clothing is performing maintenance within the vessel.

Test Reactor (TFTR), they were quicker to reach a decision and to select a site, at Princeton University. With a circular plasma and slightly smaller dimensions but higher magnetic field than JET, TFTR (Fig. 10.4) was designed for a plasma current of 2.5 MA. The Soviet Union announced tentative plans to build a large tokamak,

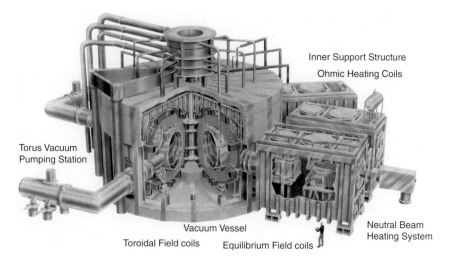

Figure 10.4 ▶ A cutaway model of the TFTR tokamak, built at Princeton, in the US, showing the magnetic field coils, the vacuum vessel, and the neutral beam heating system. It was the largest tokamak in the US fusion program until it closed down in 1998.

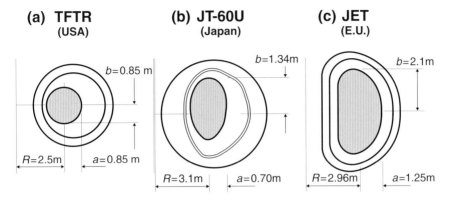

Figure 10.5 ▶ Comparison of the three large tokamaks built in the 1970s to demonstrate that fusion power increases with machine size and to get as close as possible to the conditions for net power production. (a) TFTR at Princeton in the US, (b) JT-60U at Naka, Japan, and (c) JET at Culham, UK.

designated T-20, but was beginning to fall behind in the tokamak stakes due to priorities in other areas of science, and they later settled for a smaller experiment, T-15. Japan decided to build a big tokamak, known as JT-60, specified to operate at 2.7 MA. This was the last of the big tokamaks to start operation — in April 1985. After major modifications a few years later, JT-60U (as it was then renamed) came to match JET in performance. The dimensions of the three large tokamaks — JET, TFTR, and JT-60U — are compared in Fig. 10.5. Keen competition among the three research teams provided a powerful stimulus, but there was always close collaboration and exchange of ideas and information.

10.2 Pushing to Peak Performance

The intense efforts to build TFTR came to fruition on the afternoon of Christmas Eve 1982, when, for the first time, a puff of hydrogen was injected into the torus and the magnetic coils energized. JET began operations about 6 months later, in June 1983. JET and TFTR quickly showed that they could operate reliably with currents of several million amps — much higher than previous tokamaks (Box 10.1). The gamble in taking such a big step had paid off. The pulse lengths in the new machines were much longer than had been achieved before; in JET the pulses usually lasted 10–20 seconds, but on some occasions they were increased to over 1 minute (Box 10.2). Energy confinement time increased so much compared to previous experiments that it was now quoted in units of seconds rather than milliseconds. At last, confinement time had entered the range needed for ignition; this was a major milestone in fusion research.

> **BOX 10.1** **Operating Limits**
>
> There are limits on the maximum plasma current, density, and beta (the ratio of plasma pressure to magnetic pressure) in a tokamak. Disruptions (see Box 9.1) usually determine the current and density limits; other instabilities set the beta limit.
>
> The *current limit* is conservatively taken as equivalent to $q_a > 3$, where q_a is the value of the parameter q (known as the safety factor) at the plasma edge. q_a depends on the plasma minor radius a, major radius R, vertical elongation κ, toroidal magnetic field B, as well as the plasma current I. Approximately $q_a \propto (aB/I)(a/R)\kappa$, and it can be seen that maximum current requires a torus that is vertically elongated (large κ) and "fat" (large a/R). Most reactor designs push these geometric factors as far as possible within engineering and physics constraints — typical values are $\kappa \approx 1.8$ and $a/R \approx 1/3$. Then I/aB is approximately a constant (typically about 1.4 in units of mega-amps, meters, and teslas for JET and ITER with $q_a \approx 3$). The maximum current $I \propto aB$ and increases with both plasma minor radius and toroidal field (but there is a limit to the maximum magnetic field at about 6 T in a large superconducting tokamak).
>
> The *density limit* is determined empirically by taking data from a wide range of experiments. The commonly used Greenwald limit gives a maximum for the line-average electron density (this is close to but not exactly the same as the average density) $n_G = I/\pi a^2$, where density is in units of 10^{20} m^{-3} and I is in MA. This can be rewritten in the form $n_G = (B/\pi a)(I/aB)$ to bring out the dependence on magnetic field ($n_G \propto B$) and plasma radius ($n_G \propto 1/a$), with I/aB approximately constant. It is important to note that this is an upper limit on the electron density and that the fuel ion density is reduced by the presence of ionized impurities that contribute electrons (see Box 9.6).
>
> The *beta limit* has been determined by extensive calculations of plasma stability and is in good agreement with experiments. It is usually written as average β (in %) = $\beta_N(I/aB)$. Values in the range $3 < \beta_N < 4$ are typical of present-day tokamaks, correponding to β of a few percent, but instabilities known as *neoclassical tearing modes* might impose $\beta_N \approx 2$ in ITER.

For the first few years, JET and TFTR used only the ohmic-heating effect of the plasma current. This gave temperatures rather higher than predicted from previous, smaller experiments, ranging up to 50 million degrees, but still too low for significant fusion power. The heating systems, which had been revised several times in light of results from smaller tokamaks, were added in stages and upgraded progressively over a period of years. Ultimately each of the big tokamaks had over 50 MW of heating capacity. In JET the heating was split roughly equally between neutral beams and radio frequency heating, whereas TFTR and JT-60 used a higher proportion of beam heating. Plasma temperatures reached the target of 200 million degrees; indeed TFTR got as high as 400 million. However, as in smaller tokamaks, the confinement time got worse as the heating was increased.

| BOX 10.2 | **Pulse Length and Confinement Time**

It is important to emphasize the difference between pulse length and confinement time — these quantities are sometimes confused in popular accounts of fusion. *Pulse length* is the overall duration of the plasma and usually is determined by technical limits on the magnetic field, plasma current, and plasma heating systems. In present-day experiments the pulse length usually lasts several tens of seconds, but some experiments with superconducting magnets (Chapter 11) have much longer pulses lengths. ITER, which is discussed later in this chapter, will have a pulse length of 400 s, and a fusion power plant will have a pulse length of many hours or days.

Confinement time is a measure of the average time that particles (ions and electrons) or energy spend in the plasma, and this is generally much shorter than the pulse length. The particle confinement time is usually longer than the energy confinement time because energy is lost from the plasma by thermal conduction as well as by thermal convection. The energy confinement time is an important quantity that enters in the ignition condition (Box 4.3, p. ??) and is defined as $\tau_E = E/P$, where E is the total kinetic (i.e., thermal) energy of the plasma ($E \approx 3nkTV$, where n and T are, respectively, the mean density and temperature and V is the plasma volume) and P is the total power into (or out of) the plasma. When the plasma is in steady-state thermal equilibrium, the power in (from external heating and alpha particles) equals the power out. ITER will have $\tau_E \approx 3.7$ s, and this is a modest extrapolation from present-day experiments (Box 10.4).

There were some rays of hope that confinement times could be improved. The ASDEX tokamak in Germany had found conditions known as the *H-mode* (Box 9.9) with a dramatic improvement in the energy confinement when neutral beam heating was combined with a divertor. JET had been designed without a divertor, but fortunately the D-shaped vessel allowed one to be installed. The first divertor was improvised in 1986 using the existing magnetic coils. The results were impressive and produced a record value of the $nT\tau_E$ product (density × temperature × energy confinement time, Box 4.3), twice as good as the best previously obtained. Thus encouraged, the JET team set about designing and installing a purpose-built divertor. The circular shape of TFTR made it difficult to follow the same route, but the Princeton scientists found they could get impressive results if they carefully conditioned the torus and used intense neutral beam heating. The following few years proved to be extremely successful, with both machines giving encouraging scientific results. There was intense competition to produce increasingly higher values of the $nT\tau_E$ product, and the best values came within a factor of 5 of the target needed for ignition.

JT-60 was the only one of the three big tokamaks to have been designed at the outset with a divertor, but the first version took up a lot of space and seriously restricted the maximum plasma current. JT-60 was shut down in 1989 to install a new divertor. Taking advantage of the already large dimensions of the toroidal field coils, this substantially changed the appearance of JT-60. It allowed an increase in the plasma current, up to 6 MA. The new version (renamed JT-60U) came into operation in March 1991 and began to play a major role in advanced tokamak research. As JET and TFTR turned their attention to operation with tritium plasmas, it was left to JT-60U to take the lead in pushing the performance in deuterium.

10.3 Tritium Operation

So far, fusion experiments had avoided using tritium and had operated almost exclusively in deuterium plasmas. This minimized the buildup of radioactivity of the tokamak structure due to neutron bombardment and thus made it easier to carry out maintenance and upgrades. However, it was important to gain experience with real operation in tritium plasmas, and both TFTR and JET were designed with this in mind. Both tokamaks were housed inside thick concrete walls to shield personnel and had provisions to carry out future maintenance work using tools that could be operated remotely. Encouraging results in deuterium plasmas pushed these plans forward.

In November 1991, JET became the first experiment to use a deuterium–tritium mixture. A large proportion of the JET team was in the control room for this exciting experiment. The first tests to check that the measurement systems were working properly used a concentration of only 1% tritium in deuterium. When the tritium concentration was raised to 10%, the peak fusion power rose to slightly more than 1 million watts (1 MW). At last here was a clear demonstration of controlled nuclear fusion power in a significant quantity — enough to heat hundreds of homes, though only for a second or so. The results were close to those predicted from earlier deuterium experiments, and it was calculated that about 5 MW would be produced with the optimum concentration of 50% tritium. However, it was decided to postpone this step for a few more years in order to make it easier to install the new divertor.

The TFTR team was not far behind. In November 1993 they introduced tritium into their tokamak and quickly raised the concentration to the optimum 50%. In an intensive campaign between 1993 and 1997, TFTR raised the fusion output to more than 10 MW. When JET returned to tritium experiments in 1997, improved plasmas obtained with the new divertor allowed the fusion output to reach a record value of about 16 MW, lasting for a few seconds. This was still not a self-sustaining fusion reaction because it needed about 25 MW of external heating. When various transient effects are taken into account, the calculated ratio of the nuclear fusion power to the effective input power is close to 1 — the long-sought conditions of breakeven. Fig. 10.6 shows a summary of the JET and TFTR tritium results.

In addition to the direct demonstration of fusion power, the tritium experiments in TFTR and JET gave important scientific results. Plasma heating by the

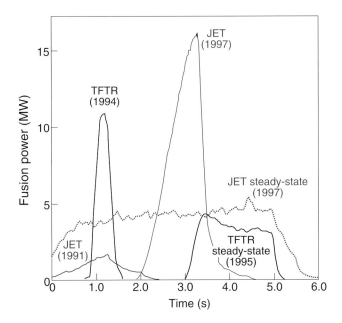

Figure 10.6 ▶ Summary of the nuclear fusion power produced with deuterium–tritium fuel in TFTR and JET during the period 1991–97. Plasmas with peak power lasting 1–2 seconds are compared with lower-power plasmas lasting for a longer time.

high-energy alpha particles from the DT reaction was measured. In JET the alpha heating was about 3 MW, compared with 20 MW external heating applied with the neutral beams. Although not yet enough to be self-sustaining, this was an encouraging demonstration of the principle of alpha particle heating. One fear that had been allayed was that some unforeseen effect might cause loss of the energetic alpha particles faster than they can heat the plasma. A welcome bonus was that energy confinement in tritium is slightly better than in deuterium. TFTR and JET also proved that tritium can be handled safely. JET tested the first large-scale plant for the supply and processing of tritium to a tokamak in a closed cycle, and the unburned tritium was reused several times in the tokamak.

10.4 Scaling to a Power Plant

The three big tokamaks, supported by the many smaller experiments, pushed magnetic confinement to conditions that are very close to those required for fusion. The results have steadily improved over time from the early Russian

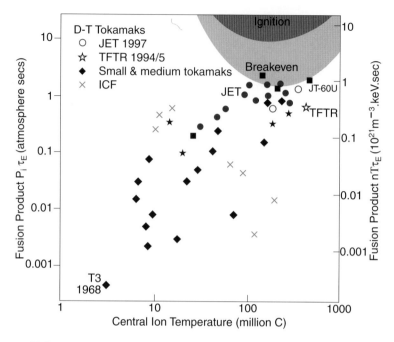

Figure 10.7 ▶ Summary of the progress toward the goal of controlled fusion. The points indicate the results actually achieved by individual experiments over many years since the Russian T3 experiment in 1968. The temperature is plotted along the horizontal axis, and the $nT\tau_E$ product is plotted along the vertical axis. JET, TFTR, and JT60U have come very close to ignition conditions, top right-hand corner. The open symbols for JET and TFTR indicate results with DT fuel. All other data are for DD plasma. Some inertial-confinement fusion results (crosses) are included for comparison.

experiment T3 in 1968 to the present day. The $nT\tau_E$ product (Fig. 10.7) has increased by more than three orders of magnitude and is within a factor of 5 of the value required for ignition. In separate experiments, temperature and density have reached or exceeded the values required for ignition. For operational convenience, most of these results are obtained in DD plasma and scaled to DT, but JET and TFTR have gone further and have values using DT plasma.

Fusion scientists extrapolate to the energy confinement times and other key parameters in ITER-sized plasmas on the basis of measurements made in many smaller experiments. It has proved very difficult to calculate the underlying particle diffusion and heat conduction from basic plasma theory. The basic processes that cause energy loss from a tokamak are known, as explained in Box 10.3, but the physics is so complex that actual loss rates cannot be calculated with sufficient accuracy to replace empirical data. In some ways the problem is similar to that of evaluating the strength of a piece of steel — this cannot be calculated from first principles with sufficient accuracy to be of practical use. When engineers need to know the strength of steel in order to build a bridge, they actually

BOX 10.3 Understanding Confinement

The simple classical theory of diffusion across a magnetic field with cylindrical symmetry (Box 5.1, p. ??) gave a diffusion coefficient of the form ρ^2/t_c, where t_c is the characteristic time between collisions and ρ is the Larmor radius in the longitudinal magnetic field. The extension to toroidal systems is known as *neoclassical theory* and takes into account the complex orbits of charged particles in a torus. In essence the relevant step length is now determined by the poloidal magnetic field. Neoclassical theory has been developed in considerable detail and is thought to be correct — but it gives values of the diffusion coefficient that are generally smaller than seen in experiments. Attempts to find refinements to the theory that remove the discrepancy have not met with success. It is now accepted that neoclassical theory describes a lower limit to diffusion that usually is overshadowed by higher losses due to some form of fine-scale turbulent fluctuations.

In spite of intensive efforts by all of the world fusion programs, it has proved extremely difficult to identify these fluctuations experimentally or to model them theoretically. This is a very complex problem, with different regimes of turbulence responsible for different aspects of plasma transport (particle diffusion, heat conduction, and so on) in different regions of the plasma — but these turbulent regimes interact with and influence each other. Thus a theory of a single regime of turbulence (and there are many such theories) is unable to provide a realistic model of the whole plasma. There has been encouraging progress in the past few years in developing a closer understanding of these processes and, in particular, in learning how to influence and control the turbulence in order to improve confinement. An example is the control of the distributions of plasma current and radial electric field in order to produce and optimize the so-called internal transport barrier.

In order to address the problem theoretically it is necessary to solve a set of highly nonlinear equations (the gyrokinetic equations) that describe plasma in nonuniform magnetic fields. It is particularly difficult to solve these equations in regimes of saturated turbulence (a problem common with all other branches of physics that deal with highly nonlinear systems). The qualitative aspects of turbulence can be modeled, but quantitative estimates of diffusion rates are much more difficult.

An alternative computational approach follows the trajectories of all the individual particles in the plasma, taking into account their equations of motion and their mutual electromagnetic interactions. This is called the *particle-in-cell* method. But the long-range nature of the electromagnetic force means that each particle interacts with virtually all the other particles in the plasma — so it is necessary to solve a very large number of coupled equations for each time step. The most powerful computers presently available are unable to carry out the calculations needed to model these highly turbulent regimes and to derive diffusion coefficients with sufficient accuracy.

---

measure the strength of small sample pieces and scale up to a larger size using their knowledge of how the strength depends on the length and cross section of the samples. In a similar way, plasma-scaling studies establish the key parameters and quantify how confinement time depends on them (Box 10.4). The empirical

> **BOX 10.4** Empirical Scaling
>
> In the absence of an adequate theoretical understanding of confinement (Box 10.3), it is necessary to rely on empirical methods to predict the performance of future experiments. Data from many tokamaks over a wide range of different conditions are analyzed using statistical methods to determine the dependence of quantities like energy confinement time on parameters like current, heating power, magnetic field, and plasma dimensions.
>
> Robert Goldston proposed one of the best-known scaling models in 1984 for tokamaks heated by neutral beams. Goldston proposed $\tau_E \propto 1/P^{0.5}$ where P is the total heating power applied to the plasma. This convenient approximate value of the exponent has proved to be remarkably resilient over the intervening two decades for a much wider range of plasma dimensions than in the original study and with other forms of heating.
>
> The ITER scaling (Figure 10.8) is based on an extensive database of confinement studies in the world's leading tokamaks. In the form in which it is usually written, it predicts strong dependence of confinement time on power ($\tau_E \propto 1/P^{0.65}$) and on plasma current ($\tau_E \propto I^{0.9}$) as well as on plasma size, magnetic field, and other parameters. However, a formulation in terms of the heating power and plasma current can be misleading because these parameters are also dependent on other parameters in the scaling expression. For example the plasma current, minor radius, and magnetic field are linked by $I/aB \approx$ constant (see Box 10.1) and P can be written in terms of τ_E, density, and temperature (see Box 10.2). A revised formulation brings out the implicit dependence on plasma temperature ($\tau_E \propto 1/T^{1.25}$) that brings the optimum temperature for DT ignition in a tokamak down to around 10 keV (see Box 4.3).
>
> The scaling for confinement time can be combined with the empirical limit for the maximum density (see Box 10.1) into an expression for the fusion triple product which shows that $nT\tau_E \propto a^2 B^3$ when quantities like I/aB and a/R are held constant. Thus it is clear that increasing the plasma radius and working at the highest possible magnetic field are the key steps in reaching ignition. There are, of course, many other parameters that need to be considered in optimizing the design of a fusion power plant, but these factors lie beyond the scope of this simplified account.

scaling of energy confinement time from present-day experiments to a power plant is shown in Fig. 10.8. There is considerable confidence in this empirical approach to confinement time scaling, which has been refined over a long period and is based on data from a very wide range of experiments.

An important parameter turns out to be the plasma current, and the extrapolation on the basis of this scaling for the confinement time indicates that a minimum of 20 million amps will be needed for ignition in a tokamak. There are limits on the maximum current (see Box 10.1) for a given plasma radius and toroidal magnetic field, and a current of 20 million amps requires a tokamak that is approximately three times bigger than JET.

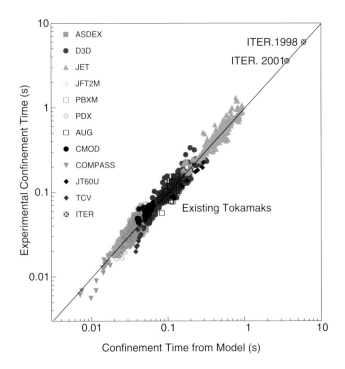

Figure 10.8 ▶ Experimental data from all the major tokamaks in the international fusion program, showing how the confinement time scales with physical parameters. Extrapolating this scaling, the confinement time is predicted to be 3.7 seconds for the present ITER design and 6 seconds for the earlier design.

10.5 The Next Step

Three major steps are foreseen to get to a commercial fusion power plant based on the tokamak concept (Fig. 10.9). The first step — to show that fusion is *scientifically feasible* — has been achieved by the three big tokamaks: TFTR, JET, and JT-60U. The *Next Step* requires an even bigger experiment (like ITER) with most of the technical features needed for a power plant (superconducting coils, tritium production; see Chapter 11) in order to show that fusion is *technically feasible*. The third step, tentatively known as *DEMO*, will be to build a full-size prototype fusion power plant producing electricity routinely and reliably to show that fusion is *commercially feasible*. Outline plans in the 1970s aimed at starting the Next Step in the 1980s and having the DEMO stage operating early in the 21st century, but various factors have combined to delay this.

Back in the late 1970s, with the construction of the big tokamaks already under way, the next task was to think about the design of the Next Step experiment. Groups in the US, the Soviet Union, the European Community, and Japan started

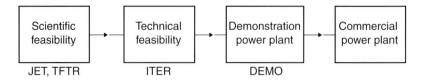

Figure 10.9 ▶ Outline of the development of the fusion power program.

to design their own versions. At the suggestion of Evgeniy Velikhov, a leading Russian fusion scientist, they came together to collaborate on a study that was known as INTOR, the International Tokamak Reactor. The INTOR study did valuable work in identifying many of the technical issues for a fusion power plant and started research programs to solve them. However, INTOR fell short on the physics specification—not surprisingly because the three big tokamaks were still only at the construction stage and it was a very large extrapolation from existing smaller tokamaks. Estimates from these small tokamaks led INTOR to be designed around a plasma current of 8 MA. But even before the study was completed it became clear that this was too small—at least 20 MA would be needed.

Not all fusion scientists have been in agreement about the strategy of this route to fusion energy based on large tokamaks. Some looked for quicker and cheaper ways to reach ignition by proposing smaller experiments to separate the fusion physics from the fusion technology. One approach seemed to be with tokamaks that operate at very high magnetic fields, allowing higher plasma current to be squeezed into a plasma with small physical size (Box 10.1 and 10.4). One problem with this approach is that these magnetic fields are too high for superconducting magnets, so this technology would not lead directly to that of a power plant. Still, the protagonists argued that if ignition could be reached this way, it would allow much of the relevant physics to be explored and would give a tremendous boost in confidence that fusion would work. Many designs for high-field-ignition experiments have been proposed along these lines, but only one has been built. This started to operate in 1991 at a laboratory near Moscow but never reached its full performance, due to a combination of technical problems and reduced funding for research in Russia. Proposals for high-field-ignition experiments in America and Europe are still being discussed but have so far failed to find support.

In addition to building ITER, research on smaller tokamaks and other confinement geometries is still an important part of the future program. Smaller experiments give the opportunity of trying a wide variety of new ideas that can, if successful, subsequently be tested at a larger scale. One line of research is into the so-called *spherical torus*, where the tokamak concept is shrunk into a much more elongated and fatter torus, with a ratio of major to minor radius of about 1. This allows it to operate at lower magnetic fields and higher plasma pressure relative to the magnetic pressure (Box 10.1). Other advantages are claimed, but this line of research is still at too early a stage to make reliable predictions about power plants.

An alternative to the tokamak might be the stellarator, and encouraging results are being obtained with a big new experiment known as the *Large Helical Device (LHD)* that started operating in Japan in 1999. Another big stellarator, W7-X, is under construction in Germany and will be completed in 2010. Conceivably these and other innovative ideas might have advantages over the tokamak and might find application in later generations of fusion power plants. But to wait for these ideas to reach the same stage of maturity as the line already explored by the big tokamaks would delay the development of fusion by many decades.

10.6 ITER

INTOR had set the scene for international collaboration in next-step experiments. At the Geneva Summit Meeting in 1985, Soviet leader Mikhail Gorbachov proposed to US President Ronald Reagan that a next-step tokamak should be designed and built as an international collaboration. The US, in consultation with Japan and the European Community, responded positively and agreed to embark on the design phase of the ITER project. The name ITER (pronounced "eater") has a dual significance — it is an acronym for International Thermonuclear Experimental Reactor and a Latin word meaning "the way." An important aspect of ITER, compared to earlier studies, was that it was regarded as a machine that would be built, though in fact the agreement did not formally make this commitment. A realistic design had to be developed and difficult problems solved, not simply put on one side.

Work on the ITER design concept started in 1988 and was followed in 1992 by a second agreement to proceed with the detailed design. This phase was carried out by a dedicated *Joint Central Team* of scientists and engineers working full time and supported by many more working part-time in so-called *Home Teams*. The lack of accord about where the team should be based and the compromise decision to split the central team between three sites (San Diego in the US, Munich in Germany, and Naka in Japan) gave early signs of future indecision at the political level.

Nonetheless the project went ahead with considerable enthusiasm, and detailed technical objectives were set. As a next-step experiment, ITER was to test and demonstrate the technologies essential for a fusion power plant, including superconducting magnets and tritium breeding, as well as exploring all the new physics of an ignited plasma. The results from tokamak experiments were now at the stage where it could be predicted that to reach ignition, ITER would require a plasma current of about 20 MA with an energy confinement time of about 6 s. ITER would be a tokamak with a divertor and a D-shaped cross section similar to JET and JT-60U with physical dimensions about three times larger than these machines. The construction was planned to take about 10 years at a cost of about $6 billion. The design had been monitored and checked by independent experts, who all agreed that ITER could and should be built.

But times had changed by 1998 when it came time to reach agreement about the construction phase. The collapse of the Soviet Union meant that a project like ITER could no longer be supported just as a showcase of East–West collaboration. Now it had to stand on its own merits as a new form of energy. Although there was much talk about environmental issues, the prospects of global warming and of future fuel shortages generated little sense of urgency to carry out the basic research to find alternative forms of energy. In fact, government funding for energy research, including fusion, had been allowed to fall, in real terms. The situation reached a crisis point in the US, where a drastic cut in research funds caused the government to pull out of the ITER collaboration. Japan, Europe, and Russia maintained their support for ITER, but the Japanese economy — seemingly impregnable only a few years earlier — was experiencing difficulties,

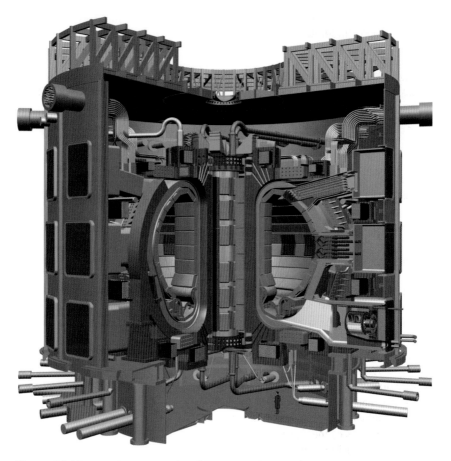

Figure 10.10 ▶ Cutaway model of the proposed International Tokamak Experimental Reactor (ITER 2001).

> **BOX 10.5** ITER — The Main Parameters
>
> The main parameters of the revised ITER design(2001) are shown in the following table.
>
> | Parameter | Symbol | Value | Units |
> |---|---|---|---|
> | Plasma major radius | R | 6.2 | m |
> | Plasma minor radius | a | 2.0 | m |
> | Plasma volume | V | 837 | m^3 |
> | Toroidal magnetic field (at plasma axis) | B | 5.3 | T |
> | Plasma current | I | 15 | MA |
> | Vertical elongation | κ | 1.86 | |
> | Average plasma/magnetic pressure | β | 2.5% | |
> | Normalized beta | β_N | 1.77 | |
> | Average electron density | n | 1×10^{20} | m^{-3} |
> | Average ion temperature | T | 8 | keV |
> | Energy confinement time | τ_E | 3.7 | s |
> | Alpha particle heating | | 82 | MW |
> | External heating power | | 40 | MW |
> | Fusion power | | 410 | MW |
> | Energy multiplication factor | Q | 10 | |

Russia was in economic crisis, and Europe did not have the determination to take the lead. If anything was to be salvaged it could be done only by accepting a delay until the political climate improved and by reducing the size in order to cut costs.

The revised design for ITER (Fig. 10.10), published in 2001 (see Box 10.5), has dimensions about 75% of those of the previous design, with a plasma current of about 15 MA. On the basis of present data, this will not be quite big enough to reach ignition — the plasma heating systems will have to be left switched on all the time — but 40 MW of heating will generate about 400 MW of fusion power. ITER will be a realistic test bed for a full-size fusion power plant in terms of technology and the physics of plasmas where the heating is dominated by alpha particles. It is possible that the reduced-size ITER could get even closer to ignition if experiments that are under way on other tokamaks prove successful.

The prospects that ITER will be built moved a step closer in 2001, when there was a Canadian offer to host ITER near Toronto on the shore of Lake Ontario. One of the attractions of the Canadian offer was that Canada's existing nuclear power plants produce tritium as a by-product. Proposals have also been put forward by Japan to host ITER at Rokkasho at the north of Honshu Island and by the European Union, which has offered Cadarache near Aix-en-Provence in the south of France. The US has come back into the project, while China and South Korea have joined the existing partners: Europe, Japan, and Russia. Many details have been agreed upon, including how the costs will be shared between the partners, and it is hoped that there will be a decision on the choice of site during 2004. If approved, construction probably will start a few years later and is estimated to take about 10 years.

▶ Chapter 11

Fusion Power Plants

11.1 Early Plans

Studies of fusion power plants based on magnetic confinement started as early as 1946, with George Thomson's patent application, which was introduced in Chapter 5. A few years later, Andrei Sakharov in the Soviet Union considered deuterium-fueled fusion and Lyman Spitzer in the US discussed systems based on stellarators. These initiatives opened the story, but it could go no further because at that time no one knew how to contain or control hot plasma. By the end of the 1960s there was growing confidence that the magnetic-confinement problem would be solved, and this prompted a renewal of activity on the subject of power plants. Bob Mills at Princeton, Fred Ribe at Los Alamos, David Rose at MIT, and Bob Carruthers and colleagues at Culham all looked into what was involved in building a fusion power plant and identified some of the problems that had to be solved. Fusion power plant design has been a feature of the magnetic-confinement fusion program ever since. A particularly significant contribution is the development of the ARIES series of studies (mainly magnetic confinement, but there is a study of an inertial-confinement power plant) by a collaboration of the University of Wisconsin, UCLA, UCSD, and other US research institutions. Some aspects of fusion power plant design are common to both magnetic- and inertial-confinement, but there are some significant differences, which will be discussed later.

11.2 Fusion Power Plant Geometry

A fusion power plant will be built up in a series of concentric layers — rather like the layers in an onion — as shown schematically in Fig. 11.1. The onion concept is particularly appropriate for an inertial-confinement power plant, where

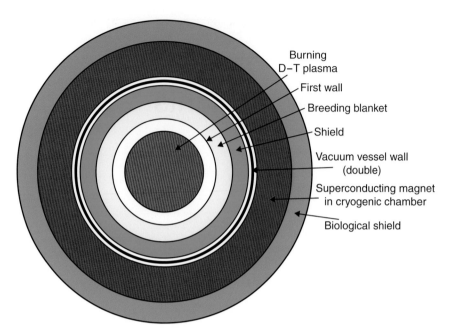

Figure 11.1 ▶ Cross section of a conceptual design for a fusion power plant. The power is extracted from the blanket and used to drive a turbine and generator, as shown in Figure 1.2. For illustration, the cross section has been shown circular, which would be the case for inertial confinement. But in a tokamak it would probably be D-shaped. No magnets are required for inertial confinement.

the topology will be spherical, but the layers in a tokamak magnetic-confinement power plant will be toroidal and D-shaped in cross section. The burning plasma forms the core, and the material surface that surrounds it is known as the *first wall*. Outside the first wall will be the *blanket*, followed by a neutron shield, the vacuum vessel, the magnetic coils (in the case of magnetic confinement), and a second shield to reduce the radiation down to the very low level required for personnel working nearby.

The average power flux at the first wall is typically several megawatts per square meter (MW m^{-2}), but it is important to remember that 80% of the power from the DT reaction is carried by neutrons, which pass through the wall without depositing much power in it. The neutrons slow down in the blanket region, depositing most of their energy within the first half meter or so, though the exact depth of the deposition depends on the composition of the breeding and structural materials. Twenty percent of the DT fusion energy is released as alpha particles and is transferred to the plasma. In an inertial-confinement power plant, all of the plasma energy is deposited on the first wall in the form of radiation and energetic particles. In a magnetic-confinement power plant, roughly half of the plasma energy is deposited in the divertor. If deposited uniformly, the typical direct heat flux onto the first wall will be less than 1 megawatt per square meter. So the first wall, the blanket, and the divertor will get hot and will require cooling by high-pressure water or helium gas, and this is the way that the fusion power will be taken out and converted into electricity. The primary coolant will pass through a heat exchanger and produce steam, which will be used to drive turbines and generators, as in a conventional power plant. Economical operation of a fusion power plant will require high power fluxes and high thermal efficiencies, thus demanding high operating temperatures. Structural materials tend to deform when operated under these conditions, and this may well be a controlling factor limiting the choice of materials.

A high power flux puts severe constraints on the mechanical and thermal design of the plasma-facing surfaces of the first wall. The problem is complicated because the surfaces facing the plasma are eroded by particles and radiation. The choice of material for these surfaces is therefore constrained by the necessity to use materials that minimize erosion and have good heat resistance. For magnetic confinement, the problem is most severe at the divertor target plates, which are subject to highly localized heat fluxes (see Box 11.2). The pulsed operation of inertial confinement poses special problems for the chamber.

The toroidal construction of a magnetic-confinement power plant traps the inner layers within the outer layers, so maintenance and repair will be difficult, and this must be taken into account at the design stage. The spherical construction makes an inertial-confinement power plant simpler to maintain, at least in principle. In both cases the blanket and first wall will have penetrations for heating and exhaust systems and for diagnostics. Some components will require regular replacement during the operating life of the plant. In both magnetic and inertial confinement, the structure surrounding the plasma will become radioactive, and "hands-on" work will not be possible after the plant has been put into operation; thus, remote-handling robots will have to carry out all of this work. Some of the techniques to do this are being developed and tested in present-day experiments, like JET.

11.3 Magnetic-Confinement Fusion

The minimum size of a magnetic-confinement fusion power plant is set by the requirement that the plasma be big enough to ignite, as discussed in Chapters 4 and 10. On the basis of the ITER scaling (Box 10.4), the minimum size for a tokamak power plant will have physical dimensions slightly larger than the present ITER design (Box 10.5), and the output will be about 3 gigawatts of thermal power, corresponding to about 1 gigawatt of electricity.

An important difference between magnetic and inertial confinement is the magnetic field. Most magnetic-confinement experiments use copper coils cooled by water. When scaled up to ITER size, copper coils would consume a large fraction of the electricity generated by the power plant, and the net output of the plant would be reduced. The answer is to use *superconducting coils* — made from special materials that, when cooled to very low temperatures, offer no resistance to an electric current. Once energized, a superconducting magnet can run continuously, and electricity is required only to run the refrigerator that keeps the coil at low temperature. Some present-day tokamaks have been constructed using superconducting coils, including TRIAM, a small Japanese tokamak that has maintained a plasma for many hours, Tore Supra at Cadarache in France, and tokamaks in Russia and China. A new superconducting tokamak is nearing completion in South Korea. There is also a large superconducting stellarator operating in Japan and one being built in Germany. Present-day superconducting materials have to be cooled to very low temperatures using liquid helium, but recent developments in superconducting technology may allow magnets to be made out of materials operating at higher temperatures, which would simplify the construction. Superconducting coils have to be shielded from the neutrons (Box 11.1) and therefore will be located behind the blanket and neutron shield. A radial thickness of blanket and shield of about 2 meters is needed to reduce the neutron flux to a satisfactory level. The absolute magnitude of the neutron flux is determined primarily by

---

BOX 11.1 Shielding the Superconducting Coils

The superconducting coils have to be placed behind the blanket and neutron shield in order to protect them from the neutrons. Some of the neutron energy is deposited as heat in the superconducting magnets and must be kept within stringent limits. The liquid helium refrigerators needed to keep the magnets at a very low temperature consume typically 500 times the energy they remove from the low-temperature coil. The refrigerators have to be powered from the power plant output, and obviously this reduces the amount of electricity that can be sold. Taking into account the efficiency of converting fusion energy as heat into electricity (typically 33%), in order to keep the consumption of power by the refrigerators to less than 3% of the plant output it is necessary to shield the superconducting coils so that they receive less than 2×10^{-5} of the neutron power flux at the first wall. This criterion will be less strict if high-temperature superconductors can be developed for large high-field coils.

A further factor to be considered is the radiation damage in the component materials of the coils. The most important are the superconductors themselves, normal conductors such as copper (these are included to stabilize the superconductor and protect it under fault conditions), and the insulators. Superconducting materials have limited ability to withstand radiation damage. Irradiation with high doses leads to a reduction in the critical current, above which the material reverts from a superconducting state to a normal resistive material. An upper limit of the total tolerable neutron dose is about 10^{22} neutrons m^{-2}, corresponding over a 20-year lifetime to an average flux of about 2×10^{13} neutrons $m^{-2} s^{-1}$. Although further work is needed to assess damage in normal conductors and insulators, the indications are that these materials should be satisfactory at the fluxes that are determined by the criteria for the superconductor itself.

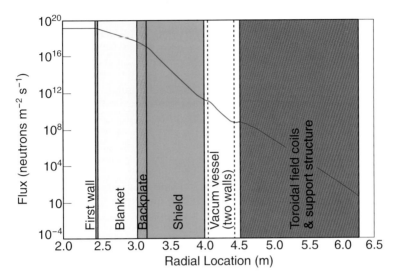

Figure 11.2 ▶ Calculated neutron flux as a function of the minor radius in a typical magnetic-confinement fusion power plant with neutron power flux of 2.2 MW m^{-2} at the first wall. The locations of the main components are indicated. The design is water cooled with a lithium lead breeder blanket, and the calculation is based on the European Environmental Assessment of Fusion Power, Model 2, 1995.

the power flux at the first wall. The neutron flux is attenuated by many orders of magnitude in passing through the blanket and neutron shield, as shown in Fig. 11.2.

Some of the fusion power transferred to the plasma from the alpha particles is radiated and deposited fairly uniformly on the first wall. If all the loss from the plasma power could be transferred uniformly to the wall, the flux of 0.5 MW m^{-2} would be relatively easy to handle, but there are various reasons why it is difficult to radiate all the power and why some power has to be removed via the divertor. Localized power loads on the divertor plate are inherently high (Box 11.2), and this is one of the most critical elements in designing a tokamak power plant. The divertor also removes the alpha particles after they diffuse out of the plasma and become neutral helium gas. This gas has to be pumped out to prevent it from building up in concentration and diluting the DT fuel. A maximum concentration of about 10% helium in the plasma is normally considered the upper limit.

11.4 Inertial-Confinement Fusion

An inertial-confinement power plant will have three major components: a *target factory* to produce target capsules, which will be compressed and heated by a *driver*

> **BOX 11.2** **Power Handling in the Divertor**
>
> The magnetic geometry constrains the plasma to follow field lines and deposits the power on the target plate in the divertor, Fig. 9.5. The width of the zone in which the power is deposited on the divertor target plate is determined by the magnetic geometry and by the inherent cross-field diffusion of the plasma—both factors tend to deposit the power in a relatively narrow annular zone. One design solution to increase the area exposed to plasma is to incline the divertor plate at a steep angle relative to the magnetic field. A second solution is to reduce the power flux to the plate by increasing the fraction that is radiated in the plasma edge. But to radiate a large amount of power it is necessary to introduce impurities. There is a limit to how much impurity can be introduced to cool the edge without cooling the center of the plasma. There are also limits to the fraction of power that can be radiated without driving the plasma unstable.

inside a *fusion chamber*, where the fusion energy will be recovered (Fig. 11.3). The blanket design, heat transfer systems, shielding, and generating plant will all be similar to those of magnetic-confinement systems. One of the big potential advantages claimed for inertial confinement is that most of the high-technology equipment is in the target factory and the driver, which can be well separated from the fusion chamber, leading to ease of maintenance. However, the target-handling systems and some optical components (including windows) have to be mounted on the target chamber, where they will be subject to blast and radiation damage. The pulsed nature of inertial-confinement fusion poses special problems for thermal and mechanical stresses in the chamber. Each capsule explodes with energy equivalent to several kilograms of high explosive, and the chamber and the blanket have to withstand the repetitive thermal and mechanical stresses and the blast from the capsule explosions. Such pulsed operation can lead to metal fatigue and additional stress above that of a continuously operating system. Some fusion chamber concepts incorporate thick liquid layers or flowing layers of granules on the inside of the chamber walls to protect them.

Present-day inertial-confinement experiments shoot less than one target per hour—in some cases only one per day. An inertial-confinement power plant will require about 10 shots every second—over 100 million each year. It will have to be fully automated and work reliably and repetitively. Targets have to be injected into the center of the target chamber at high speed, optically tracked with high precision, and then hit with the driver beams while moving. The heart of an inertial-confinement target is the spherical capsule, which contains the DT fuel, and for indirect-drive targets (see Chapter 7) this is surrounded by the hohlraum structure. Targets for present-day experiments are manufactured individually and assembled by hand using fabrication techniques that are not well suited to economical mass production. They are supported by a stalk or suspended on light

Section 11.4 Inertial-Confinement Fusion

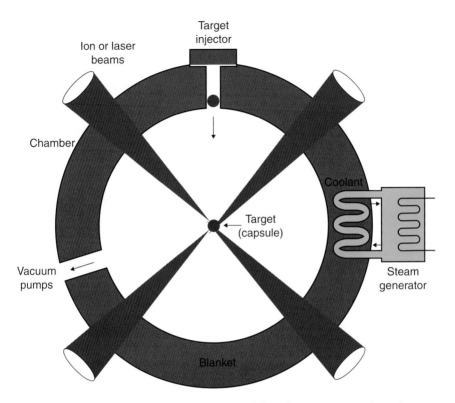

Figure 11.3 ▶ Proposed design for an inertial-confinement power plant. A target capsule is dropped into an evacuated chamber, and a pulse of laser light is fired at it to compress and heat it. The conceptual design is similar to that shown in Fig. 11.1, but there are no magnet coils. Heat extracted from the blanket will go to a conventional steam plant.

frameworks like a spiderweb. Each handmade target costs several thousand dollars, but to generate electricity at a competitive price the cost will have to fall to a few cents. This will require a highly automated target factory with a substantial development in technology.

The second big challenge to inertial fusion is to develop suitable driver systems. The driver uses electricity to produce beams of light or accelerated particles that compress and heat the target. The compressed target ignites, producing fusion energy that is converted to electricity, and some of this is needed to run the driver. The power to run the driver must be a small fraction of the total power plant output. The efficiency of the driver and target compression stages are thus critical issues, as discussed in Box 11.3.

Present-day lasers typically convert less than 1% of the electricity into target heating and compression. New types of lasers that could have higher efficiencies are being developed. The fast ignition concept (Box 7.5) separates the compression and heating, and its development could increase the efficiency by a factor of

> **BOX 11.3** **Driver Efficiency and Target Gain**
>
> As already discussed in Boxes 7.1, p. 71, and 7.2, p. 72, to break even, an ICF capsule has to yield at least enough energy to drive the next pulse. But an ICF power plant has to do rather better than just reach breakeven, and to be commercially viable it can use only a small part, say, 10%, of its own output to power the driver. So defining ϵ as the overall efficiency for converting fusion energy in the form of heat into effective capsule heating and Q as the fusion gain of the capsule (the ratio of the fusion energy released to the driver energy applied to the capsule) requires $\epsilon Q > 10$. The fusion output, deposited initially as heat in the blanket and wall of the target chamber, has to be converted to electricity at a typical conversion efficiency of 33%. The net driver efficiency (defined in terms of converting electrical energy into effective capsule drive) is $\epsilon_D \approx 3\epsilon$ and thus $\epsilon_D Q > 30$. Note that Q, as defined here, is a ratio of energies (unlike MCF, where it is a ratio of powers).
>
> A very simple estimate of Q can be obtained by comparing the fusion energy produced if the whole capsule burns with the energy needed to heat the whole capsule to the optimum temperature for ignition, $T \approx 20$ keV. The DT reaction releases 17.6 MeV, and so the fusion energy yield for a capsule with n fuel nuclei will be $\frac{1}{2}nk(17.6 \times 10^3)$ and the energy needed to heat these fuel nuclei (and their electrons) to T keV is $3nkT$. Thus $Q \approx 3000/T \approx 150$. It is important to stress that this simple calculation is included here merely to give a feel for Q. In practice it may not be necessary to heat the whole capsule to ignition temperature (especially if the fast ignition route is successful), but the burn-up will be less than 100% and the additional driver energy expended in capsule compression has to be taken into account. Calculating the capsule gain Q requires sophisticated computer models of the capsule ablation, compression, heating, and burn-up. Typical gains calculated for the NIF capsules are in the range 20–40 for indirect and direct drive and about 300 for fast ignition.
>
> A capsule gain of 30 would require driver efficiency $\epsilon_D \approx 30/Q \approx 100\%$; even a gain of 300 would require driver efficiency of 10%. Presently there are no ICF driver systems that approach this efficiency. The NIF lasers will require 350 MJ of electricity to generate 1.8 MJ of ultraviolet light—corresponding to $\epsilon_D \approx 0.5\%$ for direct drive. With a hohlraum, the ultraviolet light converts to about 350 kJ of X-rays—corresponding to $\epsilon_D \approx 0.1\%$ for indirect drive.

2 or 3. A parallel approach is to develop alternative types of drivers, including ion accelerators and X-ray sources, with higher efficiency.

A second important requirement for the driver system is that it have an adequate repetition rate and durability. This is a big problem for the lasers — the big neodymium glass lasers that are presently used in experiments can be fired only a few times each day, and their optical components are frequently damaged and need to be replaced. Accelerators should have a much faster repetition rate, although it seems unlikely that these accelerators will be able to get sufficiently uniform illumination of the target for direct heating, and it will therefore be necessary to use the indirect-drive approach (Chapter 7).

11.5 Tritium Breeding

A requirement common to both magnetic- and inertial-confinement power plants based on the DT reaction is the need to breed tritium fuel. There are small quantities of tritium available as a by-product from fission reactors (especially the Canadian CANDU reactors) which can be used to start up a fusion power program; thereafter each fusion plant will have to breed its own supplies. The basic principles of the reaction between neutrons and lithium in a layer surrounding the reacting plasma were discussed in Box 4.2. This layer, known as the *blanket*, is discussed in more detail in Boxes 11.4 and 11.5. The tritium will be extracted and fed back as fuel. Lithium is widely available as a mineral in the Earth's crust. Reserves have

---

BOX 11.4 Tritium Breeding

It is necessary to breed tritium efficiently in a fusion power plant, obtaining at least one tritium atom for every one consumed. There is only one neutron produced by each DT reaction, and so it is necessary to make sure that, on average, each neutron breeds at least one tritium nucleus. There are inevitable losses due to absorption of neutrons in the structure of the power plant and the fact that it is not possible to surround the plasma with a complete blanket of lithium — the toroidal geometry of most magnetic-confinement systems restricts the amount of blanket that can be used for breeding on the inboard side of the torus.

However, each DT neutron starts with an energy of 14.1 MeV, so in principle it may undergo the endothermic ^{7}Li reaction, producing a secondary neutron as well as a tritium nucleus. The secondary neutron can go on to breed more tritium with either the ^{7}Li or the ^{6}Li reactions. There are other nuclear reactions that can increase the number of neutrons, such as that with beryllium,

$$^9\text{Be} + n \rightarrow {}^4\text{He} + {}^4\text{He} + 2n$$

Because of the losses by absorption, calculation of the tritium breeding rate is complicated and needs to take into account the detailed mechanical structure and materials of the fusion power plant, the form in which the lithium breeding material is contained (Box 11.5), the type of coolant used in the blanket, and the fraction of the blanket that can be devoted to the breeder. Account must be taken of components that reduce the space available for the breeding blanket: plasma heating, diagnostics, and divertor systems in magnetic-confinement, and the capsule injection and driver systems in inertial-confinement. In general it should be easier to achieve tritium self-sufficiency with inertial confinement as compared to magnetic confinement because of the simpler geometry and fewer penetrations. A breeding ratio slightly greater than unity (say, 1.01) will be sufficient in principle for self-sufficiency if the tritium is recycled promptly. Ratios between 1.05 and 1.2 have been calculated for typical blanket designs. Testing these designs and confirming the predictions is an important task for the next-step experiments.

> **BOX 11.5** **Choice of the Chemical Form of Lithium**
>
> Various chemical forms have been considered for the lithium in the blanket. Possible lithium compounds are lithium/lead or lithium/tin alloys, lithium oxide (Li_2O), lithium orthosilicate (Li_4SiO_4), and a lithium/fluorine/beryllium mixture (Li_2BeF_4) known as *Flibe*. Lithium and lithium/lead offer the highest breeding ratios without the use of a neutron multiplier such as beryllium. Lead in fact acts as a neutron multiplier due to the $^{208}Pb(n, 2n)$ reaction. Metallic lithium can be used in principle. Its disadvantage in liquid metallic form is that it is very reactive chemically, and it also is easily set on fire in the presence of air or water in the case of an accident. There are also problems in pumping conducting metals at high rates across the magnetic field.

been estimated to be equivalent to a 30,000-year supply at present world total energy consumption. Even larger quantities of lithium could be extracted from seawater.

We have discussed the DT fusion reaction almost exclusively up to now. DT is the favorite candidate fuel for a fusion power plant because it has the largest reaction rate of all the fusion reactions and burns at the lowest temperature. The DT cycle has two principal disadvantages: (1) It produces neutrons, which require shielding and will damage and activate the structure; (2) the need to breed tritium requires the extra complexity, cost, and radial space for the lithium blanket. There are alternative fusion fuel cycles (see Box 11.6) that might avoid some of these problems, but these fuels are much more difficult to burn than DT. Developing fusion power plants to burn any fuel other than DT is far in the future of fusion energy.

11.6 Radiation Damage and Shielding

Fusion neutrons interact with the atoms of the walls, blanket, and other structures surrounding the plasma. They undergo nuclear reactions and scatter from atoms, transferring energy and momentum to them. Neutrons have a number of deleterious effects. Firstly, they damage the structural materials. Secondly, they cause the structure to become radioactive, which requires the material to be carefully recycled or disposed of as waste at the end of the power plant's life. However, the activity of materials in fusion power plants is confined to the structural materials, since the waste product of the fusion reaction is helium.

Radiation damage processes (Box 11.7) have been studied in considerable detail in fission reactors, providing a good basis for assessing the problems in fusion power plants. The fusion neutron spectrum consists of a primary component

> **BOX 11.6** **Alternative Fuels**
>
> There are possible alternative fusion reactions starting with deuterium:
>
> $$D + D \rightarrow p(3.02 \text{ MeV}) + T(1.01 \text{ MeV})$$
>
> $$D + D \rightarrow n(2.45 \text{ MeV}) + {}^3He(0.82 \text{ MeV})$$
>
> $$D + {}^3He \rightarrow p(14.68 \text{ MeV}) + {}^4He(3.67 \text{ MeV}).$$
>
> The two DD reactions have equal probability. The T and ^{3}He that is produced by these DD reactions would burn with more D *in situ*, adding to the fusion power and enhancing the overall reaction. However, the DD reaction rate is lower than that of DT and requires a much higher temperature — around 700 million degrees. Such a high temperature causes very high power losses from the plasma by types of radiation (known as *bremsstrahlung* and *synchrotron radiation*) that are not generally a problem with DT. The net effect is that the ignition conditions for DD are much more demanding than for DT and far exceed the capability of any known magnetic-confinement system. Starting with DD fuel (which is readily available) would avoid the need to breed tritium, but burned in this way would not substantially reduce the number of neutrons as compared to DT.
>
> Neutrons would be avoided in principle by the D^{3}He reaction. This has a higher reaction rate than DD but still falls below DT, requires high temperatures similar to DD, and thus suffers from the radiation losses. To burn D^{3}He fuel in a tokamak would require an energy confinement that is a factor 3–4 times better than present scaling (Box 10.4), with density and plasma pressure that are three to four times beyond the present limits (Box 10.1), and with very tight limits on impurities. The neutron problem is not really resolved — the D^{3}He reaction does not produce any neutrons itself, but there will be significant numbers of neutrons from parasitic DD reactions. There is a major problem obtaining fuel because ^{3}He does not occur in worthwhile amounts on Earth. Although the Moon's surface contains ^{3}He due to bombardment of particles from the solar wind, the concentration is extremely dilute, and it would require enormous mining effort to extract.
>
> There are other fusion reactions that avoid neutron production altogether:
>
> $$^3He + {}^3He \rightarrow 2p + {}^4He + 12.86 \text{ MeV}$$
>
> $$p + {}^{11}Be \rightarrow 3 \times {}^3He + 8.7 \text{ MeV}.$$
>
> However, the reaction rates for these reactions are so small that there is little prospect that they will yield economical fusion power.

of 14 MeV neutrons from the DT reaction together with a spectrum of lower-energy neutrons resulting from scattering. The damage caused by fusion neutrons is expected to be more severe than that from fission neutrons because the spectrum of fusion neutrons extends to higher energies. These higher-energy neutrons

| BOX 11.7 | Radiation Damage |

When an energetic neutron collides with an atom of the first wall or blanket structure, it can knock the atom out of its normal position in the lattice. The displaced atom may come to rest at an interstitial position in the lattice, leaving a vacancy at its original lattice site (Fig. 11.4). In fact the displaced atom may have enough energy to displace other atoms before coming to rest, and so the damage usually occurs in cascades. The damage is quantified in terms of the average number of *displacements per atom (dpa)* experienced during the working life. In places like the first wall, where the neutron flux is highest, the damage rate is expected to reach the order of hundreds of dpa. At these levels of damage the strength of the material will be reduced significantly, and some of the wall components will have to be renewed several times during the lifetime of the power plant. The neutron damage can be reduced in principle by reducing the neutron flux — but this requires a larger structure for the same output power and increases the capital costs. Thus there has to be a careful optimization of the size of the plant to strike a balance between the capital and maintenance costs.

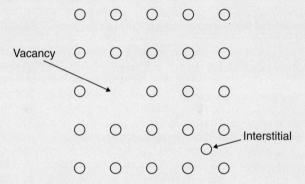

Figure 11.4 ▶ Schematic of the formation of interstitials and vacancies in a lattice structure. Hydrogen and helium atoms are also introduced into the lattice due to (n, p) and (n, α) reactions.

A neutron can also undergo a nuclear reaction with a lattice atom leaving a transmuted atom (or atoms) in place of the original. Usually the new atom is radioactive, and this is the main source of radioactivity in a fusion power plant. The neutron can also eject a proton or an alpha particle from the target nucleus, and these reactions are referred to as (n, p) and (n, α) reactions. A typical (n, p) reaction is

$$^{56}\text{Fe} + n \rightarrow p + {}^{56}\text{Mn}$$

The (n, α) reactions are particularly important for the 14 MeV neutrons from the DT fusion reaction. The protons and alpha particles produced in these reactions pick up electrons and form hydrogen and helium atoms in the lattice. Individual atoms tend to coalesce, forming bubbles of gas in the lattice, and can be deleterious to the structural strength. Further damage to the lattice can be produced by energetic recoil of the reaction products.

cause reactions that deposit helium in the solid lattice, and this has a marked effect on the behavior of materials under irradiation. To fully characterize the materials for a fusion power plant requires the construction of a powerful test facility with neutrons in the appropriate energy range.

11.7 Low-Activation Materials

One of the most important advantages of fusion relative to fission is that there will be no highly radioactive waste from the fusion fuel cycle. The structure will become radioactive by exposure to neutrons, but by careful choice of the materials the radioactivity of a fusion power plant will fall to a very low value in less than 100 years, as shown in Fig. 11.5 — this is also discussed in Chapter 12. The blanket, the first wall, and (in the case of magnetic confinement) the divertor systems will have a finite operational life and must be maintained and replaced by robots. Initial levels of radioactivity of material removed from service and the rate of decay of the various radioactive isotopes will dictate the acceptable storage and disposal methods and the possibility of recycling used components. Recycling materials, which would reduce the amount of radioactive waste, may be a possibility. To achieve low activation it is necessary to choose the structural materials carefully.

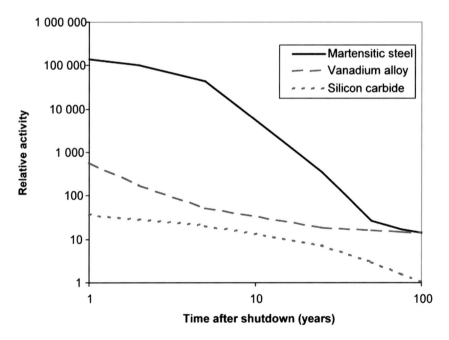

Figure 11.5 ▶ Calculated activity of a low-activation martensitic ferritic steel, vanadium alloy, and silicon carbide as a function of time from the end of power plant operation.

An important source of activation comes from small traces of impurities in the structural materials rather than from their main constituents, and careful control of quality and material purity will be critical to achieving the low-activation goal.

Although the successful development of suitable low-activation structural materials will be a formidable challenge, they would provide a significant improvement relative to presently available materials like austenitic stainless steel. Three groups of materials are under consideration as low-activation structural materials (Box 11.8). At present the low-activation martensitic steels are best understood, and they are the most probable candidates for a fusion power plant

---

BOX 11.8 Low-Activation Materials

There are three groups of materials with potential as low-activation structural materials. The first group is known as *martensitic* steels — these are used extensively in high-temperature applications throughout the world, including fission reactors. The database on properties in a radiation environment is extensive. The upper operational temperature limit, based on high-temperature strength properties, is usually estimated to be 550°C. The main constituents of martensitic steels, iron and chromium, have relatively good activation properties, and the main source of activation comes from some of the minor ingredients, like niobium. Low-activation versions can be developed by reformulating the composition (for example, replacing niobium by vanadium or titanium and replacing molybdenum by tungsten). These reformulated steels have properties very similar (and some cases superior) to those of the existing alloys. Further studies are needed on the effects of helium deposition in the lattice.

Alloys of *vanadium* form the second group of materials, and the radiation-induced characteristics are being studied specifically for the fusion materials program. The elements vanadium, titanium, chromium, and silicon, which make up most vanadium alloys, all qualify as low activation. Some vanadium alloys have been studied in the US, but it has been found that they become brittle below 400°C and above 650°C leaving a rather narrow operating range.

The third material under study is *silicon carbide* (SiC), a material that is being developed for aerospace applications. Interest in the use of silicon carbide composites stems not only from the low-activation properties but also from the mechanical strength at very high temperatures ($\sim$1000°C). This might make possible a high-temperature direct-cycle helium-cooled design. However, under irradiation there are indications that swelling of the silicon fiber and decrease of strength can occur at temperatures in the range 800–900°C. In addition to developing composites that have adequate structural strength, there are other technical problems to be solved. Methods must be developed that will seal the material and reduce its permeability to gases to acceptably low figures. Methods of joining SiC to itself and to metals must be developed. The design methodology for use of SiC composites in very large structures, at high temperatures with complex loads, and in a radiation environment is a formidable challenge. It is recognized that development of suitable materials is going to be of comparable difficulty to the solution of the plasma confinement problem.

Section 11.7 **Low-Activation Materials**

in the medium term. However, development of suitable silicon carbide composites would have the advantages of higher power plant operating temperatures, and hence higher thermal efficiencies, together with the very low activation rates for maintenance.

In considering the various engineering and scientific problems of a fusion power plant, the conclusion of a great many detailed studies and independent reviews is that no insuperable problems exist. It is within our abilities to construct such a device, but it will require a sustained effort of research and development of the technology. How long this will take depends on the urgency with which we need the results — but typically this development would take about 20 years. The important questions, whether we really need fusion energy and whether it will be economical, are discussed in the next chapter.

▶ Chapter 12

Why We Need Fusion Energy

Previous chapters have explained the basic principles of fusion, how fusion works in the Sun and the stars, and how it is being developed as a source of energy on Earth. After many years of difficult research, terrestrial fusion energy has been shown to be scientifically feasible. The next step, as discussed in the previous chapter, will be to develop the necessary technology and to build a prototype power plant. However, even if fusion energy is scientifically and technically feasible, it is also important to show that it will be economical and safe and will not damage the environment; it must be of benefit to mankind. To address these questions and attempt to answer them, we need to look beyond fusion, at the world requirements for energy today and in the future and at the various ways in which these needs might be provided.

12.1 World Energy Needs

Readers of this book who live in one of the industrially developed countries depend on an adequate supply of energy to support their way of life. They use energy to heat and cool their homes; to produce, transport, store, and cook their food; to travel to work, to go on holiday, and to visit friends; and to manufacture the vast range of goods that are taken for granted. Their life quickly grinds to a halt whenever an electrical power outage or a shortage of gasoline interrupts the supply of energy.

During the course of the 20th century, the abundance of energy has dramatically changed the way people in the industrially developed countries live and work. Energy has eliminated almost all of the labor-intensive tasks in farming and in factories that our fathers and grandfathers toiled over—reducing the quality of their lives and in many cases damaging their health. But the situation is vastly different in the developing countries, where billions of people have barely enough energy to survive, let alone enough to increase their living standards. Today, each person in the developing world uses less than one-tenth as much energy as those in one of the industrialized countries, and 2 billion people — one-third of the world's population — live without access to any electricity.

Over the last hundred years the world's commercial output and population have increased more rapidly than ever before, and the overall annual energy consumption has risen more than 10-fold. Most of the growth in commercial output and energy consumption has been in the industrialized nations, whereas most of the growth in population has been in the developing world. It took many thousands of years for the population of the world to reach 1 billion (from the last glacial maximum in 20,000 BC to 1820 AD), just a hundred years to double to 2 billion (in 1925), and then only 75 years to treble to today's population of 6 billion. Assuming that no major catastrophe intervenes, the world population is predicted to reach 8 billion in 2025 and 10–12 billion by the end of the 21st century. If people in the developing world are to achieve prosperity, their energy needs will have to be met. Finding ways of reducing our demands for energy and making more efficient use of it are important steps to reduce the overall demand — indeed just as important as finding new sources of energy. However, even if we assume that people in the industrialized world will drastically reduce the amount of energy they use, say, to one-half the present level, bringing the rest of the world up to this same level and coping with the expected growth in population will still require a big increase in world energy supply.

Many studies have been made of the increase of world population and the demand for energy, and all reach broadly similar conclusions. The examples shown in Fig. 12.1 are taken from a report entitled *Global Energy Perspectives to 2050 and Beyond*, produced by the World Energy Council in collaboration with the International Institute of Applied Systems. The figures up to 1995 are actual values, and the predictions of the future growth up to 2100 are based on a range of assumptions. For example, if one assumes that the world economy continues to grow rapidly and that no control is imposed on energy consumption, the energy demand in 2100 is predicted to be about four times its present level. A more conservative scenario, where economic growth and the use of energy is reduced by strict ecological constraints, leads to an estimated doubling of demand compared to today. Such predictions inevitably have large uncertainties, but the simple message is clear — the world will use more and more energy. By the end of the 21st century the world will be using as much energy *each 10 or 20 years* as humankind has used *in total* from the dawn of civilization up to the present day. We have to consider what reserves and what types of fuels will be available to meet these demands for energy.

12.2 The Choice of Fuels

The energy we use today comes in various forms; some are more convenient than others to a particular purpose, such as heating or transport. Wood and fuels such

Section 12.2 The Choice of Fuels

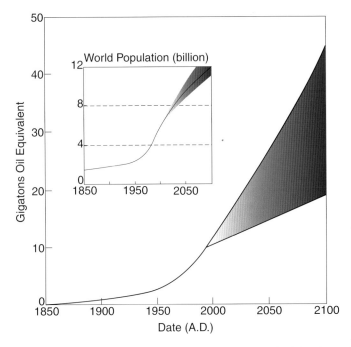

Figure 12.1 ▶ World primary energy consumption up to 1995, with estimated extrapolations up to 2100, for a range of scenarios. Inset is the estimate of world population growth, on which the energy growth is based. The spread in the predictions is determined by the range of assumptions in the models used, from an aggressive growth policy to a very ecologically restricted one. The units of energy are 10^9 tons of oil equivalent.

as animal dung and agricultural waste were once the only significant sources of energy and remain so even today in many developing countries. The exploitation of coal initiated the industrial revolution in Europe and America, and oil and gas now provide the basis for our present-day lifestyle. Coal, oil, and gas together account for about 80% of world energy consumption (Fig. 12.2), but the resources of these fossil fuels are limited and will not last forever. There are proven reserves of oil to last for between 20 and 40 years at present rates of consumption. It is widely assumed that there are much larger reserves of oil just waiting to be discovered — a common view is that the oil companies do not find it worthwhile to explore new oil fields beyond that time horizon. But some recent studies reach a much more pessimistic outlook. To keep pace with the ever-increasing demand, new oil fields would have to be found in the future at a faster rate than ever before — but most of the suitable regions of the Earth have already been explored intensively. Almost all the oil in the world may have been discovered already (but not quite all the gas), and we may already be close to having used up one-half of the world reserves of these fuels. If this scenario proves to be true, serious oil shortages and price rises will begin in the next decade. There are large untapped reserves

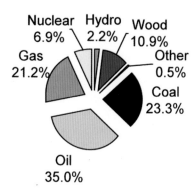

Figure 12.2 ▶ Global energy consumption in the year 2001. This includes primary energy used for generating electricity. The sector marked "wood" includes all combustible renewables and wastes; "other" includes geothermal, solar, and wind energy.

of very heavy oil in shale and so-called oil sands, but it is not known if these can be exploited in ways that are economically viable and ecologically acceptable. World reserves of coal are enormous by comparison and estimated to be sufficient to last for 200 years at present rates of use. One attraction of coal is that it is well distributed geographically, compared to oil and gas. Most of the world's remaining oil is in the Middle East — making it very vulnerable to political factors and price fluctuations.

However, long before the reserves run out there will be serious environmental problems due to the burning of so much fossil fuel. There are therefore strong arguments against allowing the use of fossil fuel — and in particular coal — to increase on the scale that would be needed to meet future energy demands. Burning a ton of coal releases into the atmosphere about 3.5 tons of carbon dioxide — and in a single year we burn nearly 5 billion tons of coal. But coal is only part of the story; when oil and gas are taken into account we release about 24 billion tons of carbon dioxide into the atmosphere each year — roughly 4 tons per person — and the amount is growing year by year. Carbon dioxide causes the atmosphere to act like a greenhouse, and a small increase in the concentration makes the Earth get slightly hotter. A small increase in the Earth's temperature has a dramatic effect on the weather and the sea level. This is global warming — probably the most devastating environmental problem that human beings have created for themselves and the toughest to solve. The process and the effects are not in doubt — all that is left to question is by how much the Earth will heat up and how long we have before drastic action is needed.

Burning fossil fuel causes other forms of damage to the environment, including acid rain, which is already affecting many rivers, lakes, and forests, and pollution, which damages people's health. Coal is particularly bad, because it contains many impurities and also because it releases more greenhouse gas per unit of energy

Section 12.2 The Choice of Fuels

than oil or gas. Burning the massive reserves of coal would be a sensible solution to future energy requirements only if new technology can be developed to burn coal without pollution and without releasing carbon dioxide into the atmosphere. A short-term response to these problems in western Europe has been to switch some electricity generation from coal to gas. But this is not a long-term solution; there is still pollution, gas supplies are limited, and at best it merely puts these problems on hold for a few more years. Burning oil and gas on a massive scale to generate electricity is also a waste of valuable resources that should be conserved for other purposes.

Electricity is an important and flexible form of energy because it is clean to use and can power a wide range of sophisticated devices like computers. About 13% of the world's energy is used in the form of electricity, but it has to be generated from a more basic form of energy. Worldwide, about 65% of electricity is generated by burning fossil fuels, about 17% from nuclear fission, and the rest mainly by hydroelectric plants. Nuclear energy is clean and produces no global warming. A few decades ago there seemed to be no doubt that this would be the energy source of the future, becoming the major source of electricity generation. France generates most of its electricity, about 80%, with nuclear fission, and Japan about 30%. Both countries have achieved higher-than-average rates of reduction in carbon dioxide emissions. The US is top of the league in terms of total nuclear energy generation, though this corresponds to only about 20% of its electricity production. However, public concern over nuclear energy and relatively low oil prices have combined to slow dramatically the construction of nuclear power plants. This is a difficult debate, driven as much by emotion as by hard fact, and it is not intended to discuss here whether adequately secure depositories for the waste from fission reactors can convince public opinion, or whether the risks from fission reactors eventually will come to be seen as more acceptable than those from global warming.

Hydroelectricity is seen as a clean source of energy — though it is not without risks or damage to the environment. It accounts for about 17% of electricity worldwide. However, most of the suitable rivers are already dammed, and there is no possibility that this source can be expanded sufficiently to meet future demands. All other forms of renewable energy — wind, tidal, and solar — taken together presently contribute less than 2% of the electricity that is generated worldwide. Although the energy in sunshine, wind, waves, and tides is enormous, there are many difficulties in harnessing these sources economically and integrating them into a reliable supply network. Those living in northern climes are very much aware that there are many days when the Sun does not shine and even those when the wind does not blow. Electricity is very difficult and expensive to store — indeed at the moment the only cost-effective method of storing it is to run a hydro plant in reverse, pumping water from a low reservoir to one at a higher level. But this adds to the costs and must be taken into account when evaluating intermittent sources like solar and wind energy. Even though these sources will be developed to meet a much greater proportion of the world's future energy requirements than they do at present, they will never be able to satisfy the total demand. New energy options must be developed — systems that are optimally safe, environmentally friendly, and economical. One long-term prospect that would further cut the cycle of pollution would be to use electricity generated by environmentally friendly sources to manufacture hydrogen. Hydrogen causes no pollution when burned and could replace fossil fuels in transport.

12.3 The Environmental Impact of Fusion Energy

One of the things that raise public concern about anything connected with the word *nuclear* is safety — the fear that a nuclear reactor might get out of control and explode or melt down. Fusion has a significant advantage compared to fission in this respect — a fusion power plant is intrinsically safe; it cannot explode or "run away." Unlike a fission power plant, which contains a large quantity of uranium or plutonium fuel, enough to keep it going for many years, a fusion power plant contains only a very small amount of deuterium and tritium fuel. Typically there is about 1 gram — enough to keep the reaction going for only a few seconds. If fuel is not continually replaced, the fusion reaction goes out and stays out.

A second safety consideration is the radioactive waste. A fission reactor produces two types of radioactive waste. The most difficult to process and store are the waste products that result from the fission fuel cycle — these are intensely radioactive substances that need to be separated from the unused fuel and then stored safely for tens of thousands of years. The fusion fuel cycle produces none of these radioactive waste products — the waste product is helium gas, which is not toxic or radioactive. The tritium fuel itself is radioactive, but it decays relatively quickly (the half-life is 12.3 years). And in any case all the tritium fuel that is produced will be recycled quickly and burned in the power plant. An important safety feature is that there need be no shipments of radioactive fuels into or out of the fusion plant. The raw materials necessary for the fusion fuel, lithium and water, are completely nonradioactive. There is enough lithium to last for at least tens of thousands of years and enough deuterium in the oceans to be truly inexhaustible. Moreover, these raw materials are widely distributed, making it impossible for any one country to corner the market. More advanced types of fusion may be developed in the very long term to burn only deuterium.

The second source of waste from a nuclear power plant is the structure of the reactor — this is made radioactive by the neutrons emitted during the nuclear reactions. Fusion and fission are broadly similar in this respect, and some component of the structure of a fusion power plant will become intensely radioactive. Components that have to be removed for repair or replacement will have to be handled using remote robots and stored inside massive concrete shields. However, the lifetime of these structural wastes is much shorter than that of fission fuel waste products. At the end of its working life, a fusion power plant will have to

Section 12.3 **The Environmental Impact of Fusion Energy**

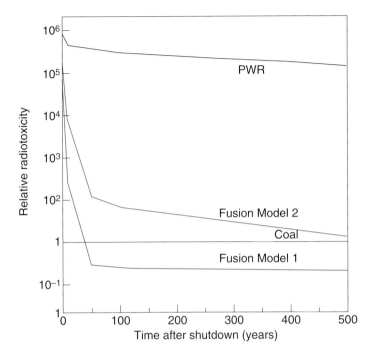

Figure 12.3 ▶ Comparison of the potential ingestion radiotoxicity of three power sources, a pressurized water (PWR) fission reactor, two models of fusion power plants, and a coal-fired power plant, all of the same electrical power output, plotted as a function of time after the end of operation.

be shielded for only 100 years before it can be completely demolished. Moreover this radioactivity can be reduced by careful choice of the construction materials — research is already under way to develop advanced steels and other materials, as discussed in Chapter 11. A comparison of the radioactivity of a fusion power plant at the end of its life with comparable data from fission and coal power plants is shown in Fig. 12.3.

With careful design and choice of materials, the level of radioactivity left by a fusion power plant after it has been closed down for about 100 years could be comparable with that left by a coal-fired power plant. Radioactivity from a coal-fired power plant is at first rather surprising. Of course burning coal does not produce any new radioactivity, but coal contains uranium (as well as many other toxic elements) which is released into the environment when coal is burned. Although the concentration of uranium in coal is relatively small, the total quantities are surprisingly large. About 3.5 million tons of coal have to be burned to produce 1 gigawatt-year of electricity (the requirement of a typical industrial city), and this contains over 5 tons of uranium. This is in fact more uranium that would be used to supply the electricity from a nuclear fission power plant. Some of the uranium escapes into the air, but most is left in the ash, which is buried in landfill sites.

12.4 The Cost of Fusion Energy

Any new source of energy, such as fusion, has to show not only that it will be clean, safe, and environmentally friendly but also that it will be competitive with other forms of energy in terms of cost. Projections using standard energy-forecasting models show that the cost of electricity generated by fusion could be competitive when it becomes available toward the middle of the 21st century. Different groups in the US, Japan, and Europe have made these projections, and there is general agreement in their results.

Of course making estimates so far ahead carries big uncertainties, and they are different for each type of fuel, so comparison on a level playing field is difficult. In the case of fossil fuels, we have a good starting point for the calculation because we know how much it costs to build and operate a coal- or gas-fired power plant today. The big uncertainties are in the future cost of the fuel (this is a major component in the cost of the electricity that is produced from fossil fuel) and the cost to the environment. Prices for oil and gas vary between countries, making comparison difficult. After substantial price rises in the 1970s, oil was relatively cheap for about 20 years, but prices started to rise substantially again in 1999. Predictions of future prices are bound to be uncertain, but prices will inevitably rise substantially as demand increases and reserves are diminished. Other factors that have to be taken into account in making long-term forecasts include environmental concerns, which might restrict or tax the use of certain types of fuel or impose more stringent pollution standards, which would push up the costs of burning these fuels. At the present time, consumers of electricity generated from fossil fuels do not pay the cost of the resulting damage to the environment or to public health. Estimates of these *external* costs are difficult because the damage is widespread and hard to quantify and affects countries other than those who cause the pollution. Nevertheless they really need to be considered in making an objective comparison between different types of fuel. Some experts calculate that the true cost of generating electricity from coal could be as much as six times higher if environmental costs were taken into account (Fig. 12.4). Advanced technologies may be developed to reduce the emission of greenhouse gas and other pollutants—but these will increase the costs of building and operating the power plants.

Making predictions about the future costs of renewable sources like wind, solar, and tidal energy is equally difficult. At the present time, electricity generated by these methods is generally much more expensive than that generated by fossil fuels, and, except for special locations, they need a government subsidy to be viable. There are also big variations from place to place. Relative costs will fall as these technologies are developed, and they are expected to become increasingly competitive as fossil fuel costs rise. Wind and solar energy are intermittent, and the cost of energy storage will need to be taken into account for these systems to be used as major sources of energy supply.

Predicting the future costs of fusion energy brings in different uncertainties. One thing we can be sure about is that fuel costs for fusion will not be important and, moreover, will be stable. In a fusion power plant the fuel will contribute less than

Section 12.4 **The Cost of Fusion Energy**

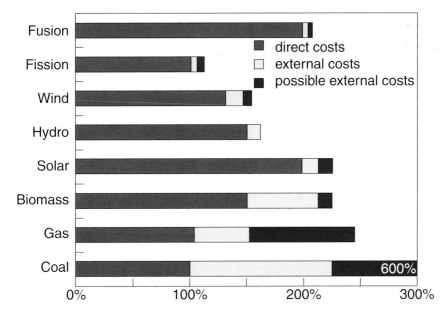

Figure 12.4 ▶ A comparison of the relative costs of electricity generated by various fuels, based on projections for around the year 2050. A range of values is indicated for each energy source, but the uncertainties are different for each fuel, as discussed in the text.

1% to the cost — a few grams of fusion fuel will produce as much energy as tens of tons of coal, oil, or gas. The dominant part of the cost of electricity generated by fusion comes from the initial capital investment in the construction of the power plant and the maintenance costs of replacing components during its working life. Many parts of a fusion power plant — buildings, turbines, and generators — will be the same as in other types of power plants, and the costs for these are well known. The plasma confinement systems will be technically complex and therefore relatively expensive to build and replace. Recent studies of potential designs based on tokamaks predict that the optimum size would generate around 1000 MW of electricity at a cost that will be competitive with other fuels. Important factors in the cost of fusion electricity are the reliability of the plant and the efficiency with which it can be operated.

When the different issues of fuel resources, costs, and damage to the environment are looked at in perspective, it is clear we will not have very many options to supply future energy requirements. Sometime during the next century we will have to stop using fossil fuels, either when the reserves run out or when they become too damaging to the environment. Renewable energy sources like wind and solar energy will play an increasingly important role, but these cannot satisfy all requirements, and we will need sources of centralized energy like fusion. An independent evaluation in 1996 by the European Commission concluded that on the basis of the progress achieved so far, the objective of a commercial fusion power plant appears to be "a demanding but reasonably achievable goal." It will probably take at least 30 years to reach this stage. If fusion is to be ready when needed, it is important that we do not delay too long before starting to build the prototype.

Epilogue

At the time when this book goes to press, there is still no decision on the site for ITER. The European Union and Japan have made strong offers to host ITER and neither appears willing to give up its claim. The most recent meeting of delegations from China, the European Union, Japan, the Republic of Korea, the Russian Federation, and the United States took place in Vienna in June 2004 and stressed the urgency of reaching a rapid resolution of the choice of the ITER site so as to move forward to implementation of ITER in a framework of international collaboration. Bilateral discussions between the European Union and Japan to resolve the deadlock are considering a broader approach to realizing fusion energy that could include international collaboration on other fusion facilities as well as ITER. This might lead to the construction of a facility known as IFMIF for testing materials exposed to a high flux of neutrons of similar energy range to those in a fusion power plant. On the one hand, it is very encouraging that Europe and Japan see ITER as being so important that both want to host it and neither will give way — but on the other hand, it would be a serious setback for fusion energy if the issue is not resolved.

The National Ignition Facility (NIF) Project has tested a single laser beam line and produced 10.4 kilojoules (kJ) of ultraviolet laser light, setting a world record for laser performance. The laser beam performance meets the primary criteria for beam energy, beam output, uniformity, beam-to-beam timing, and delivery of shaped pulses required for NIF. This demonstrated performance is equivalent to 2 million Joules (MJ) in 192 beams, exceeding the specified design requirement of 1.8 MJ.

The impending shortage of oil due to increase in global demand and the consequent increase in its cost, discussed in Chapter 12, is becoming more obvious all the time. The world's oil production is expected to reach its maximum output within the next decade and thereafter begin to decline. It is clear that the development of alternative sources of energy, such as fusion, must not be delayed too long.

Units

Scientific Notation

The subject of fusion requires that we discuss very large numbers for some things, such as the density and temperature of the hot plasma, and very small numbers for other things, like the pulse length of lasers. For example, a typical plasma temperature is 100 million degrees Celsius, and a typical plasma density is 100 million million million (1 followed by 20 zeros) particles per cubic meter. At the other extreme, the pulse length of the fastest lasers is measured in terms of one-millionth of a millionth of a second or even one-thousandth of a millionth of a millionth of a second.

Very large or very small numbers are inconvenient to write in this manner, and scientists and engineers have devised a notation in which the number of zeros in large numbers is expressed as an *exponent* above the 10. Written in this way, a temperature of 100 million degrees would be 10^8 degrees, and a density of 100 million million million would be 10^{20} particles per cubic meter. A simple extension of this idea makes it easy to express very small numbers as negative exponents — thus one-tenth is written 10^{-1}, one-thousandth 10^{-3}, and one-millionth is 10^{-6}. Some of these subdivisions have names — thus a thousandth is known as *milli* and a millionth as *micro*.

This notation will already be familiar to many readers, but for others it may be a bit off-putting. In general we have tried to spell out all the units in the main text, even though this gets a bit tedious in places, but we use scientific notation in the boxes. The following table shows the equivalence between the different ways of expressing these quantities and gives examples of the ranges of time, mass, and energy that will be encountered.

Units

Units are the measure of a particular quantity. We are all familiar with some units used in everyday life, such as degrees Celsius for temperature, cubic meters

In words	Prefix	Time (s)	Mass (g)	Energy (J)	Scientific notation
One-thousandth of a millionth of a millionth	femto-	fs			10^{-15}
One-millionth of a millionth	pico-	ps			10^{-12}
One-thousandth of a millionth	nano-	ns			10^{-9}
One-millionth	micro-	μs	μg		10^{-6}
One-thousandth	milli-	ms	mg		10^{-3}
One		s	g	J	1
One thousand	kilo-		kg	kJ	10^{3}
One million	mega-			MJ	10^{6}
One thousand million (one billion in the United States)	giga-			GJ	10^{9}
One million million	tera-			TJ	10^{12}

for volume, seconds for time, and amperes for electrical current. Each of these units has a precise scientific definition. The units with which we are principally concerned in nuclear fusion are temperature, density, time, and energy.

Temperature is usually measured in degrees Celsius — on this scale water freezes at 0°C and boils at 100°C. For scientific purposes the Kelvin scale is commonly used — absolute zero on the Kelvin scale is minus 273°C. The difference between degrees Kelvin and degrees Celsius can be ignored on the scale of temperatures for fusion — these range up to hundreds of millions (10^8) of degrees Celsius. Plasma temperatures are frequently expressed in units of *electron volts*. For comparison, one electron volt (1 eV) is equal to 1.1605×10^4 degrees Celsius — roughly 10,000 (10^4) degrees Celsius. So temperatures in the range of millions of degrees Celsius are usually expressed in terms of thousands of electron volts (keV), and 100 million degrees Celsius is approximately 10 keV.

Density in everyday use is usually expressed in terms of mass per unit volume — we think of density as kilograms per cubic meter (or perhaps still as pounds per cubic feet). The same units could be used for plasmas — but it is more convenient to think in terms of the number of particles per unit volume rather than the mass per unit volume. Densities in a magnetically confined plasma are typically in the range from 1 million million million (10^{18}) to 1000 million million million (10^{21}) particles per cubic meter. Inertial fusion densities are much larger and are generally expressed in terms of grams or kilograms per cubic meter.

Time is of particular importance in fusion experiments. For magnetic confinement of plasma, the confinement times have now reached about 1 second, a length of time we all feel familiar with. In inertial confinement the times are much shorter, typically less than 1 billionth (10^{-9}) of a second.

Energy is a term that is used in a general sense — as in "fusion energy" — and as a specific scientific quantity. Energy is measured in units of joules (J). However, energies of individual particles are usually expressed in terms of the voltage through which the particle has been accelerated and 1 electron volt (eV) = 1.6022×10^{-19} joules (this is known as Boltzmann's constant, k). The total kinetic energy of a plasma is the sum of the energies of all the ions and electrons. Temperature is a measure of the average energy, and each ion and electron has average energy equal to $(3/2)T$ (this is because a particle can move in three directions and has average energy equal to $\frac{1}{2}T$ in each direction). So if there are n ions and n electrons per cubic meter, the total plasma energy is equal to $3nkT$.

Power is the rate at which energy is used, and it is measured in units of watts (W). One watt is equal to 1 joule per second (1 W = 1 J/s). We generally use power in discussing the output of an electrical power plant — a typical home would require a power supply on the scale of several kilowatts (kW), and a typical modern large power plant has a capacity of 1 or 2 gigawatts (GW).

Glossary

ablation: The evaporation or erosion of a solid surface, usually by very high power heating.

alpha particle: A helium nucleus; emitted in radioactive decay of heavy elements, and the stable nonradioactive end product of many fusion reactions.

alternative fuels: Fusion fuels other than deuterium and tritium. They have lower probability of undergoing a fusion reaction and require much higher temperature, but may produce fewer neutrons.

arcing: An electrical discharge between two surfaces in atmosphere or vacuum, driven by an electric potential. An arc can also occur between a plasma and an adjacent surface due to the potential that naturally builds up between the two. Arcing can lead to erosion of surfaces and injection of impurities into a plasma.

ARIES: A series of design studies for fusion power plants by a consortium of US laboratories, including University of California at San Diego, University of Wisconsin, Argonne National Laboratory, Plasma Science and Fusion Center, MIT, Princeton Plasma Physics Lab, Renselaer Polytechnic Institute, General Atomics, and the Boeing Company.

ASDEX and **ASDEX-U:** Tokamaks constructed at the Max Planck Institut fur Plasma Physik, near Munich, Germany.

atom: The smallest unit of a chemical element; when it is subdivided, the parts no longer have the properties of a chemical element.

atomic mass: A, the mass of an atom relative to the standard of the carbon atom, defined as 12 units.

atomic number: Z, the number of protons in an atom.

beta: The ratio of plasma pressure to magnetic pressure – usually denoted by the symbol β.

"Big Bang": The name of the theory now widely accepted as explaining the origin of the universe.

blanket: An annular region surrounding the plasma in a fusion power plant in which neutrons would interact with lithium to generate tritium and would be slowed down to extract their energy as heat.

Boltzmann's constant: $k = 1.38 \times 10^{-23}$ J/°K or 1.6×10^{-16} J/keV converts a temperature (in °K or keV) into energy (J).

breakeven: When the power released by the nuclear fusion reactions is equal to the power used to heat the plasma.

breeding cycle: The reactions that enable tritium to be produced in a fusion power plant. A neutron released from the DT fusion reaction reacts with lithium in the breeding blanket around the confined plasma. This results in the production of a new tritium atom which can be recycled and burned in the plasma. A breeding ratio of slightly greater than unity is possible.

bremsstrahlung: Electromagnetic radiation emitted by a plasma due to the acceleration of electrons in the electric field of other charged particles.

capsule: Small pellet of fusion fuel, usually frozen DT enclosed in a plastic shell, for inertial-confinement fusion.

carbon cycle: A chain of nuclear reactions involving carbon as a catalyst, which fuses four hydrogen nuclei into one helium nucleus, with a resulting release of energy. This cycle is important in stars hotter than the Sun.

catalyst: A substance that participates in a reaction but is left in its original state at the end.

charge exchange: A process whereby an ion and an atom exchange an electron. It is a loss mechanism when fast plasma ions are neutralized and escape from the plasma, and a plasma-forming or -heating mechanism when fast atoms are injected into the plasma.

classical super: An early design of a hydrogen bomb, devised by Edward Teller, in which fusion reactions were supposed to be induced by the high temperature created by the explosion of an adjacent fission bomb. This design was shown to be impractical.

"cold fusion": Attempts made to obtain fusion at or near room temperature. Techniques using electrolysis are generally agreed not to be substantiated. Muon-catalyzed fusion has been demonstrated scientifically but is unlikely to be able to produce net energy.

conductivity: The ability of a substance to conduct heat or electricity. Plasmas are good conductors of electricity.

confinement time: The characteristic time of energy (or particle) loss from a fusion plasma. In a plasma that is in thermal equilibrium, the energy confinement time is defined as the total energy content (MJ) divided by the total power loss (MW).

convection: The transfer of heat by motion of the hotter parts of a fluid to the colder parts.

Cosmic Microwave Background Radiation (CMBR): Microwave radiation that was created in the early life of the universe, about 380,000 years after the "Big Bang".

cosmic rays: High-energy particles or nuclei that reach the Earth from outer space.

critical density: The density above which an electromagnetic wave cannot propagate through a plasma because the wave is *cut off* and reflected or absorbed.

critical mass: The mass of fissile material above which it will spontaneously undergo a nuclear fission explosion. It occurs when the material is large enough that the neutrons released can set up a chain reaction within the solid.

critical surface: The position in a plasma corresponding to the critical density.

cross section: An effective area that describes the probability that a particular collision or reaction will occur.

current limit: The maximum current allowed in a tokamak plasma. This is determined mainly by the magnetic field and plasma dimensions.

cyclotron radiation: Radiation emitted by the plasma electrons or ions due to their cyclotron motion in the magnetic field (sometimes also called *synchrotron radiation*).

density limit: An empirical limit for the maximum stable density against disruptions in a magnetic-confinement system.

deuterium: An isotope of hydrogen with one proton and one neutron in the nucleus.

diagnostics: A term used in fusion research for the many different techniques of measuring the properties of a plasma, such as density, temperature, and confinement time.

DIII-D: Tokamak at the General Atomics laboratory, San Diego, CA.

direct drive: The method of heating an inertial confinement capsule in which the energy from a laser is focused directly onto the surface of the capsule. Many laser beams have to be used to ensure uniform heating.

displacements per atom: A measure of the amount of radiation damage induced in a solid due to neutron or ion irradiation.

disruption: A gross instability of a magnetically confined plasma that causes an abrupt temperature drop and the termination of the plasma current.

divertor: A modification of the basic toroidal magnetic-confinement system that deflects magnetic-field lines at the periphery so that particle and energy fluxes can be controlled.

DOE: Department of Energy. The US governmental department that oversees fusion energy research.

driver: The means used to provide compression and heating in an inertial-confinement experiment. Up to the present, lasers have been normally used as drivers. However, energetic ion beams are also being considered.

electrolysis: The passing of electrical current through an electrolyte, causing ions in the electrolyte to flow to the anode or cathode, depending on their charge.

electrolyte: A solution in which electricity can flow.

electron: A particle with one unit of negative electric charge, 1/1836 the mass of a proton, which is not affected by the strong nuclear force.

electron cyclotron resonance heating: A technique for heating plasmas by applying electromagnetic radiation at the cyclotron frequency of the electrons. The frequency depends on the strength of the magnetic field and is typically 50–200 GHz.

Glossary

electron volt: The energy gained by an electron when accelerated through a potential difference of 1 volt. Used as a unit of plasma temperature (1 eV corresponds to 11,600°C).

element: An atom with unique chemical properties characterized by a specified number of protons in its nucleus.

emission line: Sharp energy peak in an electromagnetic spectrum, caused by a transition from one discrete energy level to a lower one.

energy level: A quantum state of a system, such as an atom, having a well-defined energy.

ELM: Edge-localized mode — a relaxation instability of the steep edge density gradient in the H-mode of a tokamak.

EURATOM: One of the three European communities that merged into the European Union, seated in Brussels, Belgium; responsible for coordinating fusion research in Europe.

evaporation: The release of material from a surface when heated to high temperature, leading to surface erosion and injection of impurities into a plasma.

excited state: The state of an atom or molecule that has a higher energy than its ground state.

fast ignition: An approach to ICF, using an ultra short-pulse, high-power laser to ignite the plasma after compression has been achieved by other means.

first wall: The solid wall directly facing the plasma in a plasma confinement experiment or a fusion power plant.

flibe: A chemical compound containing fluorine, lithium, and beryllium, that has been proposed as a possible compound for tritium breeding in a fusion power plant.

fusion: In the context of this book, nuclear fusion, the process whereby light nuclei combine to form a heavier nucleus, with the release of energy.

fusion power plant: See *nuclear fusion power plant*.

GAMMA 10: A large magnetic mirror machine built in Tsukuba, Japan.

Goldston scaling: Empirical scaling for the tokamak energy confinement time.

Greenwald limit: Empirical limit for the maximum stable density in a tokamak.

half-life: The time required for a radioactive element to decay to half its original activity.

heavy-ion driver: A particle accelerator creating intense beams of heavy ions to compress an ICF pellet.

heavy water: Water in which ordinary hydrogen has been replaced by the heavier hydrogen isotope, deuterium.

helical winding: The twisted magnetic coils that generate the rotational transform in a stellarator.

hertz: A measure of frequency with units of inverse seconds.

high Z: Materials and impurities like tungsten with large atomic number.

H-mode: A regime of improved confinement in tokamaks characterized by steep density and temperature gradients at the plasma edge.

hohlraum: A cavity or can that contains an ICF capsule. The primary driver, laser, or ion beam, shines onto the walls of the hohlraum, creating intense X-rays, which then compress the pellet more uniformly than the primary driver could.

hydrogen: The lightest of all elements. It consists of one proton and one electron. Hydrogen is sometimes used to refer to the three isotopes that have only one proton: hydrogen (protium), deuterium, and tritium.

hydrogen bomb, H-bomb: An atomic bomb based on fusion reactions. Deuterium and tritium undergo fusion when they are subjected to heat and pressure generated by the explosion of a fission bomb trigger.

ignition: In magnetic-confinement, the condition where the alpha particle heating (20% of the fusion energy output) balances the energy losses; in inertial-confinement, the point when the core a compressed fuel capsule starts to burn.

implosion: The inward movement of an ICF pellet caused by the pressure from the rapid ablation of the outer surface by the driver.

impurities: Ions of elements other than the D–T fuel in the plasma.

indirect drive: A method of heating an inertial confinement capsule in which the energy from a laser is directed onto the inside surface of a hohlraum (or cavity) in which the capsule is contained. The hohlraum surface is heated, producing X-rays that bounce around the inside of the hohlraum, striking the capsule from all directions and smoothing out any irregularities in the original laser beams.

inertial confinement fusion (ICF): By heating and compressing a plasma very rapidly, it is possible to reach high temperatures before the plasma has time to expand. Under the right conditions it is possible to reach sufficiently high temperatures and densities for nuclear fusion to occur.

instabilities: Unstable perturbations of the plasma equilibrium. Small-scale instabilities usually lead to a decrease in the plasma-confinement time; large-scale instabilities can result in the abrupt termination of the plasma (see *disruption*).

INTOR: A conceptual design study of an International Tokamak Reactor between 1978 and 1988.

ion: An atom that has more or fewer electrons than the number of protons in its nucleus and that therefore has a net electric charge. Molecular ions also occur.

ion cyclotron resonance heating (ICRH): A heating method using radio frequencies tuned to the gyro-resonance of a plasma ion species in the confining magnetic field.

ionization: The process whereby an atom loses one or more of its electrons. This can occur at high temperatures or by collision with another particle, typically an electron.

isotopes: Of an element, two or more atoms having the same number of protons but a different number of neutrons.

ITER: International Thermonuclear Experimental Reactor; an international collaboration started in 1988 to design (and hopefully construct) a large tokamak to develop the physics and technology needed for a fusion power plant.

JET: Joint European Torus; the world's largest tokamak, built at Culham, UK, as a European collaboration.

JT-60: Japan Tokamak; the largest Japanese tokamak, later rebuilt as JT-60U.

keV: Kilo-electron-volt; used as a unit of plasma temperature (1 keV corresponds to 11.6 million °C).

Kurchatov Institute: The Institute of Atomic Energy in Moscow, Russia's leading center for fusion research, named after Academician Igor Kurchatov. The tokamak was invented and developed here.

Larmor radius: The radius of the orbit of an electron or ion moving in a magnetic field.

laser: A device that generates or amplifies coherent electromagnetic waves. Laser light can be produced at high intensities and focused onto very small areas. Its name is an acronym for "light amplification by stimulated emission of radiation."

Laser Megajoule Facility (LMJ): A large neodymium laser system with a planned output of 1.8 MJ being built at Bordeaux, France, and due to be completed in 2010. It is planned to use it for inertial-confinement experiments concentrating on the indirect drive approach.

Lawrence Livermore National Laboratory: United States' nuclear weapons laboratory that was the leading center for mirror machines, now mainly devoted to inertial-confinement fusion.

Lawson criterion (or condition): The condition for a fusion plasma to create more energy than required to heat it. The original version assumed for a pulsed system that the fusion energy was converted to electricity with an efficiency of 33% and used to heat the next pulse. It is more common now, for magnetic confinement, to consider the steady-state ignition condition, where the fusion fuel ignites due to self-heating by the alpha particles.

LHD: Large Helical Device of the stellerator type built at the Japanese National Institute for Fusion Science, near Nagoya.

limiter: A device to restrict the plasma aperture and to concentrate the particle and heat fluxes onto a well-defined material surface. Originally of refractory metal (tungsten or molybdenum), later usually of carbon.

Linear pinch: Also known as a Z-pinch; an open-ended system where the plasma current flows axially and generates an azimuthal magnetic field that constricts, or "pinches," the plasma.

L-mode: Denotes the normal confinement in tokamaks.

low Z impurities: Material and impurities like carbon and oxygen with a relatively low atomic number.

Los Alamos National Laboratory: US nuclear weapons laboratory also engaged in fusion research.

low activation materials: Materials used in a radiation environment which have low activation cross sections and therefore do not get very radioactive.

magnetic axis: The field line at the center of a set of nested toroidal magnetic surfaces.

magnetic-confinement fusion (MCF): The use of magnetic fields to confine a plasma for sufficient time to reach the temperatures and densities for nuclear fusion to occur.

magnetic field lines: Directions mapping out the magnetic field in space; the degree of packing of field lines indicates the relative strength of the field.

magnetic island: A region of the magnetic field where the field lines form closed loops unconnected with the rest of the field lines. Magnetic islands can be formed by the application of specially designed external coils but can also form spontaneously under some conditions.

mass spectrograph: A device for analyzing samples according to their atomic mass. The sample is ionized, accelerated in an electric field, and then passed into a magnetic field and detected at a photographic plate. The trajectory in the magnetic field is a circular path whose curvature depends on the ratio of the mass to charge of the ion. When using electrical detection, it is termed a **mass spectrometer**.

mirror machine: An open-ended system where the plasma is confined in a solenoidal field that increases in strength at the ends, the "magnetic" mirrors.

molecule: Two or more atoms held together by chemical bonds.

mu meson, or **muon:** A particle 207 times the mass of an electron, with a lifetime of 2.2×10^{-6} s. It can have positive, negative, or neutral charge. The negative muon can substitute for an electron in the hydrogen atom, forming a muonic atom.

muon-catalysed fusion: The replacement of the electron in a tritium atom by a mu meson followed by the formation of a DT molecule, resulting in a nuclear fusion reaction with the release of a neutron and an alpha particle. The mu meson is usually released after the fusion reaction, allowing it to catalyze further reactions.

NBI: Neutral beam injection for heating and refueling magnetic-confinement plasmas.

neoclassical theory: The classical theory of diffusion applied to toroidal systems takes account of the complex orbits of charged particles in toroidal magnetic fields.

neodymium glass laser: A high-power laser, used as a driver for inertial-confinement fusion, where the active lasing element is a rod or slab of glass doped with the element neodymium.

neutrino: A neutral elementary particle with very small mass but carrying energy, momentum, and spin. Neutrinos interact very weakly with matter, allowing them to travel long distances in the universe.

neutron: A neutral elementary particle with no electric charge but mass similar to that of a proton, interacting via the strong nuclear force.

neutron star: An extremely dense star composed of neutrons; believed to be formed by supernovae explosions.

NIF: National Ignition Facility, laser-driven inertial-confinement fusion experiment being built at Livermore, with an expected completion date of 2010.

NOVA: Was the largest laser device in the United States, at the Livermore National Laboratory. Now being replaced by NIF.

nuclear fission: The splitting of a heavy nucleus into two or more lighter nuclei whose combined masses are less than that of the initial nucleus. The missing mass is converted into energy, usually into the kinetic energy of the products.

nuclear force: The strong force, one of the four known fundamental forces.

nuclear fusion: The fusion of two light nuclei to form a heavier nucleus and other particles, the sum of whose masses is slightly smaller than the sum of the original nuclei. The missing mass is converted into energy, usually into the kinetic energy of the products. Because of the electric charges on the initial nuclei, a high kinetic energy is required to initiate fusion reactions.

nuclear fusion power plant: Plant where energy will be obtained from controlled thermonuclear fusion reactions. It is intended that the ITER project will be followed by a demonstration nuclear fusion power plant.

nuclear transmutation: General term used for the nuclear reactions that result in the transformation of one nucleus into another one.

nucleosynthesis: The formation of the elements, primarily in the "Big Bang" and in the stars.

nucleus: The core of an atom; it has a positive charge and most of the atomic mass but occupies only a small part of the volume.

ohmic heating: The process that heats a plasma by means of an electric current. Known as *joule heating* in other branches of physics.

OMEGA: A large laser facility used for inertial-confinement experiments at the University of Rochester, NY.

particle-in-cell (PIC) method: A theoretical technique for studying plasma behavior by following the trajectories of all the individual particles in the fields due to the other particles. Because very large numbers of particles have to be followed, it is computationally intensive.

pellet: See *capsule*.

periodic table: A method of classifying the elements, first developed by Dimitri Mendeleev in 1869. The elements are placed in rows according to atomic mass and in columns according to chemical properties.

photon: A massless uncharged particle, associated with an electromagnetic wave, that has energy and momentum and moves with the speed of light.

photosphere: The outer layer surrounding the Sun's core that emits the Sun's energy as light and heat.

pinch effect: The compression of a plasma due to the inward radial force of the azimuthal magnetic field associated with a longitudinal current in a plasma.

plasma: A hot gas in which a substantial fraction of the atoms are ionized. Most of the universe is composed of plasma, and it is often described as the "fourth state of matter."

plasma focus: A type of linear Z pinch where the plasma forms an intense hot spot near the anode, emitting copious bursts of X-rays and neutrons.

plasma frequency: The natural oscillation frequency of an electron in a plasma.

PLT: Princeton Large Torus; the first tokamak with a current larger than 1 MA.

poloidal: When a straight solenoid is bent into a torus, the azimuthal coordinate becomes the poloidal coordinate. Thus, in a toroidal pinch or tokamak, the poloidal magnetic field is the component of magnetic field generated by the toroidal current flowing in the plasma.

positron: The antiparticle of the electron; has the same mass as the electron but opposite charge.

Pressurized Water Reactor (PWR): A widely used type of fission reactor using pressurized water as a coolant.

Princeton: The Plasma Physics Laboratory (PPPL) at Princeton University is one of the leading fusion laboratories in the United States.

proton: Elementary particle with unit positive electric charge and mass similar to that of the neutron, interacting via the strong force.

pulsar: A source of regular pulses of radio waves, with intervals from milliseconds to seconds. Discovered in 1967 and since realized to be a rotating neutron star, a collapsed massive star with a beam of radiation sweeping across space each time the star rotates.

quantum mechanics: A mathematical way of describing matter and radiation at an atomic level. It explains how light and matter can behave both as particles and waves. One of its early successes was the explanation of the discrete wavelengths of light emitted by atoms.

q-value: See *safety factor*.

Q-value: The ratio of the fusion power to heating power in a fusion device; $Q = 1$ is referred to as breakeven.

radiation damage: When energetic particles — for example, ions, electrons or neutrons — enter a solid, they can displace the atoms of the solid from their original position, resulting in vacancies and interstitials. Frequently, this happens in cascades. The result is usually a change in the mechanical and electrical properties of the solid. The energetic incident particles can also cause nuclear reactions, introducing gases such as hydrogen and helium as well as other transmuted atoms into the solid. The deterioration in the mechanical properties is a problem in nuclear fission power plants and needs to be taken account of in fusion power plant design.

Rayleigh–Taylor instability: A hydrodynamic instability that can occur when two fluids of different densities are adjacent to each other, resulting in intermixing or interchange. It is a serious problem during ICF pellet implosion.

reconnection: It is convenient to visualize a magnetic field in terms of magnetic field lines — rather like lengths of string that map out the vector directions of the field. Sometimes magnetic field lines become tangled and sort themselves out by breaking and reconnecting in a more orderly arrangement.

red shift: The change in wavelength toward the red end of the spectrum of radiation coming from the stars and other objects in outer space. It is a Doppler shift due to the expansion of the universe.

RF heating: Radio frequency heating; methods of heating a plasma by means of electromagnetic waves at various plasma resonance frequencies.

RFP: Reverse field pinch; version of the toroidal pinch where the stabilizing toroidal magnetic field reverses direction on the outside of the plasma.

safety factor: A measure of the helical twist of magnetic field lines in a torus, denoted by the symbol q. The value of q at the plasma edge in a tokamak must be greater than about 3 for stability against disruptions.

Sandia National Laboratories: US nuclear weapons laboratory in Albuquerque, NM, that is engaged in fusion research on both light and heavy ion drivers for inertial-confinement fusion as well as Z-pinches.

sawteeth: Periodic relaxation oscillations of the central temperature and density of a plasma in the core of a tokamak.

scaling: Empirical relations between quantities such as the energy confinement time and various plasma parameters.

scrape-off layer: The region at the plasma edge where particles and energy flow along open field lines to the limiter or divertor.

SCYLLA: An early theta pinch at Los Alamos.

sonoluminescence: The emission of light by bubbles in a liquid excited by sound waves.

spectrograph: A device to make and record a spectrum of electromagnetic radiation.

spectrometer: A device to make and record a spectrum electronically.

spectroscopy: The measurement and analysis of spectra.

spectrum: A display of electromagnetic radiation spread out by wavelength or frequency.

spherical tokamak or torus: A toroidal configuration with an extremely small aspect ratio, where the toroidal field coils of a conventional tokamak are replaced by a straight rod.

Spitzer conductivity: The electrical conductivity of a fully ionized plasma was calculated by Lyman Spitzer, who showed that it scales as $T_e^{3/2}$ (the temperature of the electrons to the power of 3/2). A pure hydrogen plasma with $T_e = 1.5$ keV conducts electricity as well as copper at room temperature.

sputtering: An atomic process in which an energetic ion or neutral atom arriving at a solid surface ejects a surface atom, thus eroding the surface.

steady state theory: A theory that accounted for the expansion of the universe by assuming that there is continuous creation of matter. It has now been replaced by the Big Bang theory.

stellarator: A generic class of toroidal-confinement systems where rotational transform of the basic toroidal field is produced by twisting the magnetic axis or by adding an external helical field. Proposed in 1951 by Lyman Spitzer at Princeton University. The class includes the early figure-eight and helical winding (classical) stellarators at Princeton as well as torsatrons, heliacs, and modular stellarators.

sticking: The process in muon-catalyzed fusion when the mu meson is carried off by the alpha particle and is therefore unable to catalyze any further fusion reactions.

stimulated emission: In lasers, the process whereby a light beam causes excited atoms to emit light coherently, thereby amplifying the beam.

superconducting: Property of a conductor to lose its resistivity at low temperature; refers to a machine with superconducting magnetic field coils.

superconductivity: The property of certain materials that have no electrical resistance when cooled to very low temperatures.

supernova: The explosion of a large star when all the nuclear fuel has been burned. Many neutrons are produced and reactions forming heavy elements occur.

T-3: Tokamak at the Kurchatov Institute in Moscow, whose results in the late 1960s prompted a worldwide emphasis on tokamaks.

Teller-Ulam configuration: The design of the H-bomb in which the radiation from a fission bomb was used to compress and heat the fusion fuel. This design was used for most of the successful H-bombs.

TFTR: Tokamak Fusion Test Reactor; the largest US tokamak experiment, operated until 1997 at Princeton University.

thermonuclear fusion: Fusion reactions with a significant release of energy occurring in a plasma in which all the interacting particles have been heated to a uniformly high temperature.

theta pinch: A confinement system, generally open ended, where a rapidly-pulsed axial magnetic field generates a plasma current in the azimuthal (i.e., theta) direction.

Thomson scattering: Method of measuring the electron temperature and density of a plasma, based on the spectral broadening of laser light scattered from electrons.

tokamak: A toroidal-confinement system with strong stabilizing toroidal magnetic field and a poloidal field generated by a toroidal current in the plasma; proposed in 1951 by Andrei Sakharov and Igor Tamm in the Soviet Union and now the most promising of the magnetic-confinement systems.

toroidal field: In a tokamak or stellarator, the main magnetic field component, usually created by large magnetic coils wrapped around the vessel that contains the plasma.

toroidal pinch: The first toroidal-confinement scheme, where plasma current in the toroidal direction both heats the plasma and provides the poloidal (pinch) magnetic field.

transuranic elements: Elements with mass higher than that of uranium. They are all radioactive and do not occur naturally, but can be made by neutron bombardment or fusing two naturally occurring elements.

triple-alpha process: A chain of fusion reactions in which three helium nuclei (alpha particles) combine to form a carbon nucleus.

tritium: An isotope of hydrogen containing one proton and two neutrons in the nucleus. It is radioactive, with a half-life of 12.35 years.

tritium breeding: See **breeding cycle**.

turbulence: In plasmas or fluids, fluctuations that can cause enhanced transport of heat and particles.

visible light: Light to which the human eye is sensitive, i.e., in the wavelength range 390–660 nanometers.

wavelength: The distance over which a wave travels in one complete oscillation.

weak interaction: One of the four fundamental forces known in nature, the weak interaction is responsible for the process of beat decay where a neutron converts into a proton and an electron.

white dwarf: The name given to the final state of a small star (mass equal to or smaller than the sun) after it has burned all the fusion fuel.

X-rays: Electromagnetic radiation of wavelength between 0.01 and 10 nanometers.

ZETA: Zero Energy Thermonuclear Assembly; a large toroidal pinch experiment at Harwell, UK; built in 1957 and for many years the largest fusion experiment in the world.

Z-pinch: A device where a plasma is confined by the azimuthal magnetic field generated by an axial current flowing in the plasma. In a straight cylindrical pinch, z is the axial coordinate and θ is the azimuthal coordinate; in a toroidal pinch, the current is toroidal and the field poloidal.

Further Reading

General

R Herman, *Fusion: The Search for Endless Energy*, Cambridge University Press, New York, 1990; ISBN: 0521383730.

JL Bromberg, *Fusion: Science, Politics and the Invention of a New Energy Source*, MIT Press, Cambridge, MA, 1982, ISBN: 0262521067.

E Rebhan, *Heisser als das Sonnen Feuer*, R Piper Verlag, Munich, 1992, (in German).

S Eliezer, Y Eliezer, *The Fourth State of Matter: An Introduction to Plasma Science*, 2nd ed. Institute of Physics Publishing, Philadelphia, 2001, ISBN: 0750307404.

Internet Sites

There are many internet sites for fusion. Here are some of the more general and interesting ones.

http://www.jet.efda.org/

http://www.jet.efda.org/pages/fusion-basics.html

http://fusedweb.pppl.gov/CPEP/Chart.html

http://www.nuc.berkeley.edu/fusion/fusion.html

http://fusioned.gat.com/

http://www.iea.org/Textbase/publications/index.asp

Chapter 1 — What Is Nuclear Fusion?

HG Wells, *The World Set Free*, Macmillan, London, 1914.

Chapter 3 — Fusion in the Sun and Stars

JD Burchfield, *Lord Kelvin and the Age of the Earth*, University of Chicago Press, Chicago, 1975, Reprinted 1990, ISBN: 0226080439.

G Gamow, *Creation of the Universe*, London, Macmillan, 1961.

J Gribbin, M Gribbin, *Stardust*, Penguin Books, London, 2001, ISBN: 0140283781.

J Gribbon, M Rees, *The Stuff of the Universe: Dark Matter, Mankind and Anthropic Cosmology*, Penguin Books, London, 1995, ISBN: 0140248188.

SW Hawking, *A Brief History of Time: From the Big Bang to Black Holes*, Bantam Press, London, 1995, ISBN: 0553175211.

CA Ronan, *The Natural History of the Universe from the Big Bang to the End of Time*, Bantam Books, London, 1991, ISBN: 0385253273.

Chapter 4 — Man-made Fusion

TK Fowler, *The Fusion Quest*, The John Hopkins University Press, Baltimore, 1997. ISBN: 0801854563.

P-H Rebut, *L'Energie des Etoiles: La Fusion Nucleaire Controlee*, Editions Odile Jacob, Paris, 1999. ISBN: 2-7381-0752-4 (in French).

J Weise, *La fusion Nucléaire*, Presses Universitaires de France, Paris, 2003 ISBN: 2 13 053309 4 (in French).

Chapter 5 — Magnetic Confinement

CM Braams, PE Stott, *Nuclear Fusion: Half a Century of Magnetic Confinement Fusion Research*, Institute of Physics, Bristol, U.K., 2002, ISBN: 0750307056.

Chapter 6 — The Hydrogen Bomb

R Rhodes, *Dark Sun, The Making of the Hydrogen Bomb*, Simon & Schuster, 1995, ISBN: 068480400X.

Further Reading

L Arnold, *Britain and the H-Bomb*, Palgrove, U.K., 2001, ISBN: 0333947428.

TB Cochran, RS Norris, "The Development of Fusion Weapons," Encyclopaedia Britannica 29, 1998, pp. 578–580.

Chapter 7 — Inertial-Confinement Fusion

J Lindl, *Inertial Confinement Fusion: The Quest for Ignition and Energy Gain Using Indirect Drive*, Springer Verlag, New York, 1998, ISBN: 156396662X.

Chapter 8 — False Trails

F Close, *Too Hot to Handle: The Story of the Race for Cold Fusion*. W. H. Allen Publishing, London, 1990, ISBN: 1852272066.

JR Huizenga, *Cold Fusion: The Scientific Fiasco of the Century*, Oxford University Press, Oxford, U.K., ISBN: 0198558171.

Chapter 9 — Tokamaks

J Wesson, *Tokamaks*, 3rd ed., Clarendon Press, Oxford, U.K., 2003, ISBN: 0198509227.

Chapter 10 — From T3 to ITER

J Wesson, *The Science of JET*, JET Joint Undertaking, Abingdon, U.K., 2000.

Chapter 12 — Why We Need Fusion Energy

BP Statistical Review of World Energy, 52nd ed. British Petroleum, London, June 2003.

JT Houghton, *Global Warming: The Complete Briefing*, 2nd ed., Cambridge University Press, Cambridge, U.K., 1997, ISBN 0521629322; 3rd ed., June 2004.

International Energy Agency, *Energy to 2050 (2003) — Scenarios for a Sustainable Future*, ISBN 92-64-01904-9.

International Energy Agency, "World Energy Outlook," to be published November 2004.

Index

Ablation, 70–73, 80, 82, 103, 161
Alarm clock, 66
Alpha particle, 41, 44, 93, 161
Alpher, Ralph, 24
Alternative fuels, 139, 161
Alverez, Luis, 93
Ambipolar effect, 51
Arcing, 103, 161
ARIES, 129, 161
Artsimovich, Lev, xiii, xiv, 58
ASDEX, ASDEX-U, 105, 109, 110, 161
Ash, helium, 103, 106
Aston, Francis, 10–13, 26
Atkinson, Robert, 14
Atom, 9, 161
Atomic mass, 9–12, 15, 161
Atomic number, 12, 104, 162
Atoms for Peace conference, 59
Azimuthal magnetic field, 57

Basov, Nicolai, 73, 76
Becquerel, Henri, 7, 18
Beryllium bottleneck, 27
Beta limit, 116
Beta, 116, 162
Bethe, Hans, 14, 22
Big Bang, 24–26, 162
Black holes, 32
Blackman, Moses, 47
Blanket, 5, 130, 137, 162
Boltzmann's constant, 44, 50, 162

Bordeaux, 79, 86
Bradbury, Norris, 62
Brahe, Tycho, 28
Breakeven, 71–73, 83, 86, 111, 118, 136, 162
Breeding cycle, 4, 36, 137–138, 162
Bremsstrahlung, 139, 162
Bulganin, Nikolai, 55, 58

Cadarache, 127, 132
CANDU, 137
Capsule, 69–72, 77–86, 162, 171
Carbon cycle, 22, 162
Carruthers, Bob, 129
Catalyst, 22, 92–94, 162
Chadwick, James, 12
Charge exchange, 107, 162
Classical super, 62, 63, 162
Cockroft, John, 58, 59
Cold fusion, 87–94, 163
Conductivity, electrical, 40, 53, 163
Confinement time, 41, 44, 99, 102, 110, 115–117, 121–123, 163
Convection, 22, 23, 117, 163
Convective envelope, 22
Cosmic microwave background radiation (CMBR), 24–25, 163
Cosmic rays, 92, 163
Critical density, 97–98, 163
Critical mass, 61, 163
Critical surface, 73, 163

Cross section, 37–38, 163
Culham, 112, 115, 129
Current limit, 106, 116, 163
Cyclotron radiation, 102, 164
Cyclotron radius, see Larmor radius

Darwin, Charles, 17, 18
DEMO, 123
Density limit, 97–98, 116, 164
Department of Energy, see DOE
Deuterium, 4, 12, 33, 61, 65, 88, 164
Diagnosing plasma, 100–102
Diagnostics, 100–102, 164
Didenko, Yuri, 92
DIII-D tokamak, 101, 164
Direct drive, 79, 81, 136, 164
Displacements per atom, 140, 164
Disruption, 97, 98, 116, 164
Diverted magnetic field lines, 104
Divertor, 104–105, 114, 117, 164
DOE, 164
Doppler broadening, 24, 101–102
Driver, 71–72, 81–86, 133-136, 164
Dynamic hohlraum, 83, 84

Eddington, Arthur, 2, 12, 13
Edge-localized modes (ELMs), 109, 165
Effective ion charge, 104
Einstein, Albert, 2, 7, 8
Electrolysis, 34, 87–89, 164
Electrolyte, 89, 164
Electron, 9, 164
Electron cyclotron radiation, 102
Electron cyclotron resonance, 108, 165
Electron volt, 37, 42, 46, 158, 165
Elements, 8, 165
ELM, see Edge-localized modes
Emission line, 165
Energy, 7
    conservation of, 7
Energy confinement time,
       see Confinement time
Energy level, 74–75, 165
Eniwetok Atoll, 65, 66
EURATOM, 165
Evaporation, 103, 165
Excited state, 74, 165

Fast ignition, 81, 83, 165
Fermi, Enrico, 15, 61, 62
First wall, 103, 130, 133–134, 165
Fission reactor, 149, 151
Flash tube, 74, 81
Fleischmann, Martin, 87–91
Flibe (lithium/fluorine/beryllium)
    mixture, 138, 165
Frank, Charles, 92
Fuchs, Klaus, 63, 64
Fusion, 2, 165
    inertial confinement 4, 41, 45–46,
        69–85, 133–136, 167
    magnetic confinement 4, 41, 45–46,
        47–60, 95–128, 131–133
    muon-catalyzed, 92–94
    thermonuclear, 39, 175
Fusion power plant, 130, 165

GAMMA10, 57, 166
Gamow, George, 14, 25, 178
Gerstein, Sergei, 93
Gigavolt, 38
Global energy consumption, 148
Goldston scaling, 122, 166
Goldston, Robert, 122
Gorbachov, Mikhail, 125
Greenwald limit, 116, 166
Gyro radius, see Larmor radius
Gyrokinetic equations, 121

H-bomb, see Hydrogen bomb
H-mode, 109–110, 117, 166
Hahn, Otto, 3, 15
Half-life, 35, 141, 166
Hartek, Paul, 38
Harwell, 52, 58
Heavy-ion driver, 82, 136, 166
Heavy water, 87, 88, 166
Helical winding, 52, 54, 55, 166
Helium, 2, 5, 9, 19, 20, 106
    ash, 103, 106
    ^{3}He 9, 19, 34
    ^{4}He 9, 19, 35
Hertz, 166
High confinement, see H-mode
High Z, 104, 166
Hohlraum, 79–81, 166
    dynamic, 83, 84

Index

Houtermans, Fritz, 14
Huxley, Thomas, 17
Hydrogen, 2, 9, 12, 19, 166
Hydrogen bomb, 61–67, 166
Hydrostatic equilibrium, 21

ICF, *see* Fusion, inertial confinement
Ion cyclotron resonance heating (ICRH), 108, 166
Ignition, 42–46, 70, 80, 111, 167
Ignition criterion, 43
Implosion, 64, 65, 167
Impurities, 102–104, 134, 167
Indirect drive, 79, 81, 86, 134, 136, 167
Inertial-confinement fusion (ICF), *see* Fusion, inertial confinement
Inertial-confinement power plant design, 135
Instabilities, 79, 96–99, 167
Institute of Advanced Study, 62
International Thermonuclear Experimental Reactor, *see* ITER
International Tokamak Reactor, *see* INTOR
INTOR, 124, 125, 167
Ion, 11, 40, 167
  charge-to-mass ratio, 11, 108
Ion cyclotron resonance, 108
Ionization, 103, 107, 167
Ionized gas, 39
ISIS, 94
Isotopes, 12, 167
ITER, 113, 116, 117, 125–128, 131, 167
  design concept, 125
  possible locations, 127
  revised design, 127
  scaling, 122
Ivy–Mike, 65

Jeans, James, 14
JET, 112, 114–116, 118, 119, 125, 131, 168
John, Augustus (painter), 13
Joint European Torus, *see* JET
Jones, Steven, 88, 94
JT-60 and JT-60U, 115, 118, 125, 168

Kadomtsev, Boris, 99
Kepler, Johannes, 28
Kharkov, 55
Khrushchev, Nikita, 55, 58
Killer pellet, 98
Kurchatov Institute, 99, 168
Kurchatov, Igor, 55, 58

L-mode, 109, 168
Large Helical Device (LHD), 54, 132
Larmor radius, 49, 50, 121, 132, 168
Laser, 72–75, 101–102, 168
Laser Megajoule (LMJ), 83, 86, 168
Lattice structure, 140
Lavrentiev, Oleg, 55
Lawrence Livermore National Laboratory, *see* Livermore
Lawson condition, 41–44, 60, 71, 168
Lawson, John, 41, 42, 71
Lebedev Institute, 73, 75
LHD, *see* Large Helical Device
Limiter, 104, 168
Linear pinch, 57, 168, *see also* Z-pinch
Lithium, 5–7, 24, 35, 65, 81, 137
Lithium deuteride, 65
Livermore, 53, 64, 75, 76, 79, 86, 168
  NIF, 78
  NOVA, 75, 77
LMJ, *see* Laser Megajoule
Lord Kelvin, 17, 18
Lord Rayleigh, 79
Los Alamos, 15, 47, 52, 57, 61, 62, 94, 129, 169
Low activation materials, 141, 142, 169
Low confinement, *see* L-mode
Low-Z impurities, 104, 169
Lower hybrid resonance, 108

Magnetic axis, 169
Magnetic confinement, *see* Fusion, magnetic confinement
Magnetic field lines, 55–57, 169
Magnetic island, 99, 169
Magnetic mirror machine, 54, 56–57, 169
Maimon, Theodore, 73
Manhattan Project, 3, 61
Martinsitic steel, 142
Mass, 7
Mass defect, 14–15

Mass spectrograph, 11, 169
Matthöfer, Hans, 112
Maxwellian distribution, 101
Medium-Z impurities, 104
Mendeleev, Dmitri, 8
Metastable state, 74
Mills, Bob, 129
Mirror machine, see Magnetic mirror machine
Mu meson (muon), 92, 169
Muon-catalyzed fusion, 92–94, 170
Muonic atom, 92

Nagamine, Kanetada, 94
National Ignition Facility see (NIF)
Nature, 59, 75, 89
Neoclassical tearing modes, 116
Neoclassical theory, 121, 170
Neodymium glass laser, 74, 75, 79, 86, 102, 136, 170
Neutral beam heating, 106–110, 116–117, 170
Neutrino, 21, 22, 170
Neutron, 3, 12, 33–36, 170
Neutron star, 32, 170
NIF (National Ignition Facility), 78, 85, 86, 170
NOVA, 75, 77, 170
Novosibirsk conference, 60, 95, 96
Nuckolls, John, 75, 76
Nuclear fission, xv, 1, 14, 61, 148, 151, 153, 170
Nuclear force, 36, 170
Nuclear fusion, xv
Nuclear transmutation, 1
Nucleosynthesis, 26–32, 171
Nucleus, 9, 171

Oak Ridge National Laboratory, 54, 91, 92
Ohmic heating, 106, 116, 171
Oliphant, Marcus, 38
OMEGA, 75, 171
Oppenheimer, Robert, 62

Palladium, 88
Palmer, Paul, 88
Particle-in-cell method, 121, 171
Pathological science, 91
Periodic table, 8, 171

Petrasso, Richard, 90
Photon, 74, 171
Photosphere, 22, 171
Pinch, 48, 54
  linear Z-, 57
  reverse field, 54
  theta, 52, 57
Pinch effect, 47, 171
Plasma, 20, 39, 172
  current, 48, 116
  density, 41, 116
  focus, 57, 172
  frequency, 172
  heating, 106-110, 116-119
  instabilities, 51
  pressure, see Beta
PLT, 111, 172
Plutonium, 61
Poloidal direction, 52, 55, 172
Poloidal field, 95, 97
Ponomarev, Leonid, 94
Pons, Stanley, 87–91
Positron, 20, 21, 172
Potential energy, 37
Powell, Cecil, 92
Pressurized water reactor (PWR), 151, 172
Princeton, 172, 174
Prokhorov, Alexandr, 73, 76
Proton, 9, 12, 20 ,172
Pulsars, 32, 172
Pulse length, 115, 117

Q-value, 127, 136, 172
q-value, see safety factor
Quantum mechanics, 2, 14, 172

Rabi, Isidor, 62
Radiation damage, 138, 140, 173
Radiation implosion, 64
Radiative zone, 22, 23
Radiofrequency (RF) heating, 108, 173
Ramsey, William, 22
Rayleigh–Taylor instability, 79, 82, 173
Reagan, Ronald, 125
Rebut, Paul, 112, 113, 178
Reconnection, 99, 173
Red shift, 24, 173
Reverse field pinch (RFP), 54, 173

Index

Ribe, Fred, 129
Rokkasho, 127
Rose, David, 129
Rutherford Appleton Laboratory, 94
Rutherford, Ernest, 3, 9, 18

Safety factor, 98, 116, 173
Sakharov, Andrei, 66, 67, 93, 129
Saltmarsh, Michael, 92
Sandia National Laboratory, 81, 83, 84, 173
Sawteeth, 98–99, 173
Scaling, 122–123, 173
Schalow, Arthur, 73
Scientific feasibility, 123
Scrape-off layer, 105, 173
SCYLLA theta pinch, 57, 174
Shapira, Dan, 92
Shutdowns, 100
Silicon carbide, 142
Sloyka, 66
Solar envelope, 21
Sonoluminescence, 91, 174
Spectrograph, 174
Spectrometer, 174
Spectroscopy, 174
Spectrum, 174
Spherical torus, 124, 174
Spitzer conductivity, 53, 174
Spitzer Space Telescope, 53
Spitzer, Lyman, 52, 53, 104, 129
Sputtering, 103, 174
Star wars, 62
Steady-state theory, 24, 174
Stellarator, 52, 54, 55, 60, 95, 96, 129, 174
Sticking, 94, 174
Stimulated emission, 74, 174
Strassman, Fritz, 3, 15
Sudbury Neutrino Observatory, *see* SNO
Superconducting, 130–132, 174
Supernova, 27, 175
Suslick, Kenneth, 92
Synchrotron rsadiation, 139

Tabak, Max, 80
Taleyarkhan, Rusi, 91
Tamm, Igor, 67
Target factory, 133
Technical feasibility, 123

Teller, Edward, 61–63
Teller–Ulam configuration, 64, 65, 67, 175
Temperature, 4, 39, 101, 157
Teniers, David, 2
Teravolt, 38
TFTR, 114–120, 175
Thermonuclear fusion, 39
Theta pinch, 52, 57, 175
   SCYLLA, 57
Thomson scattering, 101, 175
Thomson, George, 4, 47, 48
Thomson, J. J., 4, 10
Thomson, William (Lord Kelvin), 17, 18
Thonemann, Peter, 47, 48
Tokamak, 54, 60, 95, 130, 175
   DIII-D, 101, 164
   divertor, 105
   instabilities, 98
   operation of, 100
   PLT, 111
   T-3, 120, 175
   T-4, 111
   T-20, 115
   TFR, 111
   Tore Supra, 132
Tokamak Fusion Test Reactor, *see* TFTR
Toroidal direction, 52–56
Toroidal field, 52–55, 95, 175
Toroidal pinch, 47–49, 52–55, 58–59, 175
Torus, 47–49
Townes, Charles, 73, 76
Transuranic elements, 15, 175
TRIAM, 132
Triple-alpha process, 27, 176
Tritium, 4, 12, 34, 63, 118, 176
Tritium breeding, 36, 137–138, 176
Truman, Harry, 63
Tsukuba, 57
Turbulence, 121, 176
Tyndall, John, 17

Ulam, Stanislaw, 64

Vanadium, 142
Velikhov, Evgeniy, 124
von Weizacker, Carl, 22

W7-X, 125
Wagner, Fritz, 109

Wavelength, 19, 176
Weak interaction, 19, 176
Wedgewood Benn, Anthony, 112
Wells, H. G., 3, 178
White dwarf, 27, 176
World Energy Council, 146

X-rays, 64, 79–81, 83, 86, 136, 176

Z facility, 84
Z-pinch, 57, 84, 176
Zero Energy Thermonuclear Assembly, *see* ZETA
ZETA, 58, 59, 176

COMPLEMENTARY SCIENCE SERIES

Physics in Biology and Medicine

SECOND EDITION

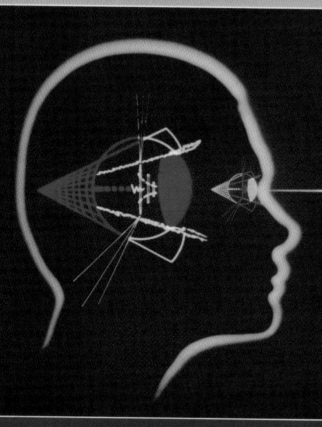

Paul Davidovits

Physics in Biology and Medicine

SECOND EDITION

Paul Davidovits *Boston College*

This is a book you should consider if you are teaching the one-semester premed course. This text could be used in two ways: 1) as a text for a one-term course in the physics of the body (without calculus) for non-physics majors in premed or allied health programs, or 2) as a supplementary text for the introductory physics course, particularly for premed students.

—RUSSELL HOBBIE
University of Minnesota, retired

There is certainly a viable market [for this book], if not as a stand-alone physics text, as a collection of problems, examples, and discussions at the boundary between physics and biology/medicine. It is very well written; it is certainly accurate; and it is pretty complete.

—DAVID CINABRO
Wayne State University

Paul Davidovits, Professor of Chemistry at Boston College, was co-awarded the prestigious year 2000 R.W. Wood prize from the Optical Society of America for his seminal work in optics. His contribution was foundational in the field of confocal microscopy (discussed herein), which allows engineers and biologists to produce optical sections through 3-D objects such as semiconductor circuits, living tissues, or a single cell. Dr. Davidovits earned his doctorate, masters, and undergraduate degrees from Columbia University. Prior to his appointment at Boston College, he was a faculty member at Yale University. He has published more than 100 papers in physical chemistry.

At one time scientists believed that a "vital force" governed the structure and organization of biological molecules. Today, most scientists realize that organisms are governed by the laws of physics on all levels.

While almost two centuries of research have found that physical laws fully apply to biology, work is far from complete. Basic questions at the atomic, molecular, and organismal levels remain unanswered. Even when typically complex molecular structure is known, function is not yet predictable. Nourishment, growth, reproduction, and communication distinguish biological matter from inorganic matter, yet these mechanisms are understood only qualitatively.

This book furthers our understanding by relating important concepts in physics to living systems. Applications of physics in biology and medicine are emphasized, with no previous knowledge of biology required. The analysis is largely quantitative, but only high-school physics and mathematics are assumed. Underlying basic physics appears in appendices. Biological systems are described in only enough detail for physical analysis.

The organization is similar to basic physics texts: solid mechanics, fluid mechanics, thermodynamics, sound, electricity, optics, and atomic and nuclear physics. A bibliography gives important sources for further reading.

www.harcourt-ap.com

ISBN 0-12-204840-7

COMPLEMENTARY SCIENCE SERIES

Introduction to Relativity

John B. Kogut

Introduction to Relativity

John B. Kogut
University of Illinois at Urbana-Champaign

This is an excellent text that covers special relativity and a first time introduction to general relativity at an accessible level. The book achieves its goals by providing a pedagogical derivation of all the important concepts from first principles and using only elementary mathematical tools. It should fill an important gap in the literature.
—MIRJAM CVETIC
University of Pennsylvania

This book is an excellent introduction to special relativity in a manner that appropriately reflects the modern view of the subject.... With this text in hand, there is little excuse for any good undergraduate in any of the sciences to graduate without a pretty good understanding of the essentials.
—CHARLES C. DYER
University of Toronto

John Kogut is professor of physics at the University of Illinois at Urbana-Champaign. His specialty is high-energy theoretical physics, in particular, the physics of quarks and gluons. He has authored more than 200 articles and reviews, including pioneering papers on the light-cone approach to field theory, the Parton model, the statistical mechanics approach to field theory, and computational methods in high energy (lattice gauge) theory. Dr. Kogut was nominated for the 1987 Nobel Peace Prize (with M. Weissman, L. Gronlund, and D. Wright) by thirty members of the U. S. House of Representatives. A former Sloan Foundation and Guggenheim Fellow, he was educated at Princeton and Stanford Universities.

This book is a unique, concise, and accessible foundation for the study of special and general relativity. It is written for anyone drawn to the rich intellectual and philosophical implications of relativity, especially undergraduate physics, engineering, and astronomy majors who may go on to study modern astrophysics, cosmology, and unified field theories.

Special relativity is developed from two basic notions: that force-free frames of reference are indistinguishable and that nature possesses a universal speed limit—the speed of light. Basic results are introduced with a minimum of algebra by constructing clocks and meter sticks so that time dilation, space contraction, and the relativity of simultaneity can be derived, explained, and illustrated. Minkowski diagrams are introduced to visualize these effects concretely. The Twin Paradox is resolved in detail. The roles of force, energy and momentum, and conservation laws in particle collisions are presented and illustrated through discussions and problems.

Although many topics of general relativity require some mathematical and physical background, Chapters 7 and 8 of the book successfully describe the core tenets at the basic level of the previous six chapters on special relativity. The Equivalence Principle is explained as the centerpiece of Einstein's theory of gravity. It states that a uniform gravitational field produces an environment that is physically indistinguishable from that in a uniformly accelerating reference frame. The gravitational red shift, the Twin Paradox as a problem of twins aging at different rates in different gravitational potentials, and the bending of light in a gravitational field are discussed and illustrated. The formulation of gravity as the theory of curved space-time is introduced at an elementary level.

The book contains a wealth of instructional and thought-provoking problems.

ISBN 0-12-417561-9

HARCOURT ACADEMIC PRESS

A Harcourt Science and Technology Company
www.harcourt-ap.com

Printed in the United States of America

COMPLEMENTARY SCIENCE SERIES

The Physical Basis of Chemistry

SECOND EDITION

Companion Multimedia Web site for this title:
http://www.princeton.edu/~chm205/Introduction_files/v3_document.htm

Warren S. Warren

The Physical Basis of Chemistry

SECOND EDITION

Warren S. Warren *Princeton University*

Praise for the first edition:

"Both [Warren's] choice of material and his style and flair of presentation are exceptionally good."
—DUDLEY HERSCHBACH
Harvard University

"Professor Warren writes clearly and carefully. His expression is at a high level but it is presented in an inviting manner to students—not condescending and not too cute."
—RICHARD N. ZARE
Stanford University

...is even more broadly applicable to the current generation of students...

"This is a great book to supplement either an advanced general chemistry course or a junior-level physical chemistry course. It could serve opposite functions in those two settings, but would work well in either. As a supplement to an introductory chemistry textbook, it would provide mathematically advanced students with additional challenge and rigor. As a supplement to a physical chemistry textbook, it would provide a bridge between the standard introductory material and the mathematically more sophisticated physical chemistry texts."
—DEBORAH HUNTLEY
Saginaw State University

Warren S. Warren, Professor of Chemistry at Princeton University, received his Ph.D. in Chemistry from U.C. Berkeley in 1980. His publications range from *Physical Review Letters* and invited papers in *Science* on his research in nuclear magnetic resonance and ultrafast laser spectroscopy to the *Journal of Chemical Education*. He received the 1982 Nobel Laureate Signature Award of the American Chemical Society and has held numerous fellowships.

Lord Ernest Rutherford, 1908 Nobel Laureate in Chemistry, put it bluntly:

Science is divided into two categories: physics and stamp collecting.

But he would have been astonished to see the transformation of biology from "stamp collecting" into molecular biology, genomics, biochemistry, and biophysics in this century. This transformation occurred only because, time and time again, fundamental advances in theoretical physics drove the development of useful new tools for chemistry. Chemists in turn learned how to synthesize and characterize ever more complex molecules, and eventually created a quantitative framework for understanding biology and medicine.

This book presents the physical, mathematical, and statistical concepts necessary for understanding the structure and function of molecules. The emphasis is placed on understanding the critical core material in quantum mechanics, thermodynamics, and spectroscopy that should be understood by any scientist or student of science. It is designed to enhance any general chemistry text by reintroducing concepts that require a little mathematical sophistication. It is also useful as a stand-alone background text for introductions to materials science, biophysics, and clinical imaging.

HARCOURT
ACADEMIC
PRESS

www.harcourt-ap.com

ISBN 0-12-735855-2

9 780127 358550

COMPLEMENTARY SCIENCE SERIES

Chemistry Connections | SECOND EDITION

The Chemical Basis of Everyday Phenomena

Kerry K. Karukstis and Gerald R. Van Hecke

Chemistry Connections

SECOND EDITION

Kerry K. Karukstis and Gerald R. Van Hecke

Harvey Mudd College

Chemistry Connections perks the reader's interest by starting off with interesting and relevant questions. Then the authors weave in essential concepts without the reader even knowing it. The references at the end of each section are EXCELLENT.
—JENNIFER SHEPHERD
Gonzaga University

This wonderful and fun book includes a large number of real-world everyday examples, organized by everyday questions, and centered in core chemical areas. Buy one, keep it on your desk as you prepare class, and use it every day to add relevant examples to all your chemistry courses.
—TRICIA A. FERRETT
Carleton College

This is a very effective way to increase students' interest in chemistry... a broad spectrum of chemical examples that are interesting and informative for a wide audience. The selected examples provide a welcome dose of reality that shows why it is both interesting and fun to learn the underlying chemical principles that surround us. The inclusion of "Other questions to consider" offers a convenient way to connect related concepts between the different examples.
—JEFFREY B. ARTERBURN
New Mexico State University

Dr. Kerry K. Karukstis, Ph.D Duke (physical chemistry), is Professor of Chemistry at Harvey Mudd College and has research interests in applications of absorbance and fluorescence spectroscopy. Her curriculum interests include lasers, biophysical applications, and experiments in physical chemistry.

Dr. Gerald R. Van Hecke, Ph.D Princeton (physical chemistry), is Professor of Chemistry at Harvey Mudd College and has research interests in liquid crystalline materials and laser scattering to measure thermodynamic properties of liquids. His curriculum interests have focused on introducing undergraduates to applications of lasers in chemistry.

ACADEMIC PRESS
An imprint of Elsevier Science
www.academicpressbooks.com

Printed in the United States of America

Chemistry Connections provides a fascinating array of examples of chemistry at work, spanning topics from the aurora, to medicine, to sticky notes. The explanations begin with the basics, followed by more detailed analyses that show why it is interesting, fun, and useful to learn the underlying chemical principles. This much-enjoyed book, now fully revised and expanded, illustrates how chemistry governs much of our everyday experience and interaction with the world around us.

Provocative and topical examples are grouped according to their chemical principles, but also linked to related topics throughout the book. **Chemistry Connections** provides both discussion and answers to the questions in both lay and technical terms with clear and concise explanation. It sheds light on some engaging aspects of chemistry that inform even professional chemists.

- Two levels of explanations: general, accessible ones highlight the chemical essence of the phenomenon; and technical ones using chemical principles provide more in-depth interpretation
- Indexing of questions according to key principles or terms enhances instructional use
- Figures and 3-D chemical structures illustrate the chemical concepts presented
- References to related World Wide Web sites for further exploration provide inexpensive and convenient access to related information.
- Color plates enhance connections between specific topics

Students taking general chemistry courses don't always appreciate the significance of principles they're learning. More general readers, put off by traditional chemistry pedagogy, are drawn in by this focus on examples from daily life. This is a suitable complementary text for any general chemistry course (for majors and non-majors) designed to acquaint students with how chemistry and science affect their lives.

ISBN 0-12-400151-3

COMPLEMENTARY SCIENCE SERIES

Fundamentals of Quantum Chemistry

SECOND EDITION

James E. House

Fundamentals of Quantum Chemistry

SECOND EDITION

James E. House   *Illinois State University*

"I wish to thank Professor J. E. House for [the first edition] which I found in the Chemistry library here at Columbia. His book is excellent. It states the basics clearly, and he teaches what a new student needs to know. When I read the author's text he seemed like my own professor because he was so generous."
— ALISON WINFIELD
Astronomy major, Columbia University

"This is an excellent book to use as an introduction to the techniques and concepts met in theoretical chemistry by undergraduate and beginning graduate students.... The organization and style of the book are such that a student would find it easy to read and follow the physical, chemical and mathematical principles."
— JIM McTAVISH
Liverpool John Moores University

"[This is a] brief 'need to know' introduction to some of the more important topics in elementary quantum mechanics. Over the years I have used two other books similar to this one, both of which were inferior to House's book."
— KURT CHRISTOFFEL
Augustana College

James E. House is Emeritus Professor of Chemistry at Illinois State University and taught at Western Kentucky University, the University of Illinois, and Illinois Wesleyan University. In addition to authoring about 150 scientific publications, he is the author of a book on Chemical Kinetics and has coauthored a book on descriptive inorganic chemistry with his wife, Katherine. He has BS and MA degrees from Southern Illinois University and a Ph.D. from the University of Illinois.

The application of quantum-mechanical principles to chemical problems has revolutionized the field of chemistry. The knowledge of quantum mechanics is now indispensable to many areas of the physical sciences. Our understanding of chemical bonding, spectral phenomena, molecular reactivities, and other fundamental chemical problems relies on the detailed behavior of electrons in atoms and molecules. Much of applied quantum mechanics is based on the treatment of several model systems such as particle in a box, harmonic oscillator, rigid rotor, barrier penetration, etc. These models are surveyed herein.

Many texts in quantum chemistry exist for the advanced student or specialist but there are few good books that provide nonspecialists with the basis of experiments and theories in their fields. This book provides a clear, readable presentation of the basic principles and application of quantum mechanical models for chemists while maintaining a level of mathematical completeness that enables the reader to follow the developments.

The second edition features:

- A new chapter on molecular orbital calculations (extended Hückel and self-consistent field) has been included which introduces the basic ideas and terminology.

- The photoelectric effect, the perturbation treatment of the helium atom, orbital symmetry and chemical reactions, and molecular term symbols are now included.

- A significant number of additional figures and minor improvements to existing figures and new exercises have been added. Answers are now provided for selected problems at the back of the book.

- The entire text has been carefully and extensively edited to increase the clarity of the presentation and to correct minor errors.

The reader needs only basic physics and calculus, but a few mathematical topics are included in sufficient detail to bring the reader up to speed with differential equations and determinants. This book is an ideal tool for self-teaching.

ELSEVIER
ACADEMIC PRESS

www.academicpressbooks.com

ISBN: 0-12-356771-8

9 780123 567710

COMPLEMENTARY SCIENCE SERIES

Earth Magnetism

A Guided Tour through Magnetic Fields

Wallace Hall Campbell

Earth Magnetism

Wallace Hall Campbell *National Oceanic and Atmospheric Administration*

From the Reviews

"This layman's non-technical description of the Earth's magnetic field is long overdue. Dr. Campbell has done an excellent job of not only describing the scientific nature of this area of science, but also puts the understanding at the level of the majority of the public."
—JOE HIRMAN
Chief Forecaster, Space Environment Forecast Center, NOAA

"This book is written by a world-renowned scientist and provides a wealth of scientific information about magnetic fields, in a way that is state-of-the-science yet fun to read. Dr. Campbell carries an unbridled enthusiasm for geomagnetism, which he is willing and able to share with scientists and non-scientists alike, and he does it with superb clarity, simplicity and practicality."
—HERBERT W. KROEHL
Secretary General, International Association for Geomagnetism and Aeronomy

Wallace Hall Campbell is guest scientist at the Solar-Terrestrial Physics Division of the National Oceanic and Atmospheric Administration in Boulder, Colorado. His current appointment follows a distinguished career as a research geophysicist within the geomagnetism group at the USGS and many years of upper atmosphere research in geomagnetism for NOAA. His research interests continue to be ionospheric currents, electrical conductivity of the Earth, geomagnetic storms, pulsations, and field applications. Dr. Campbell wrote and co-authored two major textbooks and more than a hundred publications in geomagnetism.

We are now in the years of maximum solar activity. Spectacular outbursts of particles and fields will bombard the Earth and continue at high levels for the first few years of the millennium. In this period of striking auroras and satellite-damaging magnetic storms, it is natural to wonder about the magnetic fields that guide solar particles to create such astonishing displays. This book takes the reader on a tour through the Earth's magnetic fields with a minimum of jargon and mathematical detail.

Journey with the author from historical observations of the magnetic field and magnetic applications in the modern world through the Earth's principle magnetic field to understand why there are so many magnetic pole locations. The everyday main field and quiet-time variations that are superimposed on the main field together form a baseline from which we measure the magnificent but potentially damaging magnetic storms. Learn how these spectacular field disturbances start from blasts of particles ejected from our Sun.

ISBN 0-12-158164-0

Harcourt Academic Press

A Harcourt Science and Technology Company
www.harcourt-ap.com

Printed in the United States of America

COMPLEMENTARY SCIENCE SERIES

Fusion

The Energy of the Universe

Garry McCracken • Peter Stott

Fusion THE ENERGY OF THE UNIVERSE

Garry McCracken UKAEA, Culham Division   **Peter Stott** UKAEA, Culham Division

"I read the McCracken and Stott book with the greatest interest. As they say, one can't put it down. It reads like a novel! While I'm probably biased, being a fusion researcher myself, I'd think the book would appeal to the scientifically literate public."
—PETER STANGEBY, University of Toronto

"This book is a delightful account of the role played by nuclear fusion in the universe, and of man's attempt to harness it for terrestrial electric power generation. It describes a broad spectrum of fusion nuclear physics and plasma physics topics, and to my knowledge is unique in its coverage. I found it to be very readable, and I believe it would be a valuable addition to the library of all those physicists and engineers currently working in the field of fusion energy."
—GEORGE C. VLASES, University of Washington

"This text provides a non-technical treatment of fusion, covering the sun's heat source, fusion weapons, and commercial power production. It is suitable for undergraduate or graduate students with limited science backgrounds. Any university with a fusion-related program would find it of interest. I also think it would be a good way to cover fusion in undergraduate institutions or graduate schools without a fusion program."
—JAMES BLANCHARD, University of Wisconsin

Garry McCracken gained a PhD in solid state physics from Queens University, Kingston, Ontario, Canada, but has spent most of his life as an experimental physicist working on various aspects of the magnetic confinement fusion program with the UK Atomic Energy Authority at Culham Laboratory. When the Joint European Torus (JET) Joint Undertaking was set up as a European Fusion Laboratory to build the JET experiment, he led a task agreement on the plasma boundary physics. There, his group built and installed major diagnostics on JET, and an active experimental program was pursued. In 1993 he went to the Massachusetts Institute of Technology, USA, and worked on the C-Mod tokamak in the Plasma Fusion Center. In 1996 he returned to the UK to continue his work on JET, until his retirement in 1999. He has published over 300 scientific papers, including three major reviews in the general area of plasma-surface interactions.

Peter Stott earned his PhD in theoretical and experimental plasma physics working between Manchester University and the Harwell and Culham Laboratories. He joined the UK Atomic Energy Authority at Culham Laboratory in 1966 and has spent his professional career as an experimental physicist working on magnetic confinement fusion. In 1979 he joined the JET Joint Undertaking to take charge of the design and construction of the plasma diagnostics systems, and from 1982 to 1999, he was head of JET's Experimental Division 1. From 1989 to 1999, he was coordinator for the European contribution to the design of diagnostics for the International Thermonuclear Experimental Reactor (ITER) project and was a member of the International Advisory Group. He left JET in 1999 to move to the Département des Recherches sur la Fusion Contrôlée, Cadarache, France. He has published over 200 scientific papers, edited six books on plasma diagnostics, and co-authored two books on fusion energy.

books.elsevier.com

Fusion powers the stars, and could in principle provide almost unlimited, environmentally benign, power on Earth. Harnessing fusion has proved to be a much greater scientific and technical challenge than originally hoped. In the early 1970s the great Russian physicist Lev Andreevich Artsimovich wrote that "thermonuclear [fusion] energy will be ready when mankind needs it". It looks as if he was right and that time is approaching. *This excellent book is therefore very timely.*

As early as 1920 it was suggested that fusion could be the source of energy in the stars, and the detailed mechanism was identified in 1938. It was clear by the 1940s that fusion energy could in principle be harnessed on Earth, but early optimism was soon recognized as being (in Artsimovich's words of 1962) "as unfounded as the sinner's hope of entering paradise without passing through purgatory." That purgatory involved identifying the right configuration of magnetic fields to hold a gas at over 100 million degrees Celsius (10 times hotter than the center of the Sun) away from the walls of its container. The solution of this challenging problem—which has been likened to holding a jelly with elastic bands—took a long time, but it has now been found.

Garry McCracken and Peter Stott have had distinguished careers in fusion research. Their book appears at a time when fusion's role as a potential ace of trumps in the energy pack is becoming increasingly recognized. I personally cannot imagine that sometime in the future, fusion energy will not be widely harnessed to the benefit of mankind. The question is when. This important book describes the exciting science of, the fascinating history of, and what is at stake in mankind's quest to harness the energy of the stars.

From the Foreword by Chris Llewellyn Smith

Professor Sir Chris Llewellyn Smith FRS is Director UKAEA Culham Division, Head of the Euratom/UKAEA

COMPLEMENTARY SCIENCE SERIES ORDER FORM

Save 15% when you order direct!

Ship to:

ELSEVIER

NAME _____ INSTITUTION/COMPANY _____

STREET ADDRESS _____

CITY _____ STATE/COUNTRY _____ ZIP _____

PHONE ( ) _____

EMAIL _____

I do not wish to receive information on ☐ Elsevier products ☐ third party products.

BOOKS

ISBN	AUTHOR/TITLE	LIST PRICE	DISCOUNT PRICE	QTY	TOTAL
0-12-204840-7	Davidovits: Physics in Biology and Medicine, 2nd Edition	$38.95/£23.95	$33.11/£20.36		
0-12-417561-9	Kogut: Introduction to Relativity for Physicists and Astronomers	$34.95/£20.99	$29.71/£17.84		
0-12-481851-X	McCracken & Stott: Fusion: The Energy of the Universe	$44.95/£29.99	$38.21/£25.49		
0-12-735855-2	Warren: The Physical Basis of Chemistry, 2nd Edition	$34.95/£22.99	$29.71/£19.54		
0-12-400151-3	Karukstis & Van Hecke: Chemistry Connections: The Chemical Basis of Everyday Phenomena, 2nd Edition	$34.95/£24.99	$29.71/£21.24		
0-12-356771-8	House: Fundamentals of Quantum Chemistry, 2nd Edition	$46.95/£29.99	$39.91/£25.49		
0-12-106051-9	Blinder: Introduction to Quantum Mechanics in Chemistry, Materials Science, and Biology	$47.95/£29.95	$40.76/£25.46		
0-12-158164-0	Campbell: Earth Magnetism: A Guided Tour Through Magnetic Fields	$31.95/£19.95	$27.16/£16.96		

METHOD OF PAYMENT

☐ CHECK ☐ VISA ☐ MC ☐ AMEX ☐ DISCOVER

CARD # _____

SIGNATURE _____

Sales, VAT or GST taxes will be added where applicable

BOOKS SUBTOTAL _____

SALES TAX ____ %
(Customer's local tax rate)

SHIPPING* _____

TOTAL _____

SHIPPING CHARGES IN THE AMERICAS, ASIA, & AUSTRALIA

- All U.S. & Canadian prepaid orders are shipped free of charge. Billed orders sent Ground: $7 on 1st item, & $8 on 2 or more
- ____ flat rate (inc. Alaska & Hawaii)
- Please call for express charges
- Outside North America, please call for charges

SHIPPING CHARGES IN EUROPE, MIDDLE EAST & AFRICA

- Prepaid orders are shipped free of charge when ordered from our European office

TO PLACE ORDERS OR FOR PRICING AND PRODUCT INQUIRES, CONTACT:

IN THE AMERICAS, ASIA, & AUSTRALIA:
Elsevier, Order Fulfillment, DM77541
11830 Westline Industrial Drive, St. Louis, MO 63146
Toll free in North America 1-800-545-2522 • Fax: 1-800-535-9935
Outside North America 1-314-453-7010 • 1-314-453-7095

IN EUROPE, MIDDLE EAST, & AFRICA:
Books Customer Service
Linacre House, Jordan Hill
Oxford, OX2 8DP, UK
Tel +44 1865 474110 • Fax: +44 1865 474111
E-MAIL: eurobkinfo@elsevier.com

AEG/LB/FR- 25291- 01/04